Artificial Intelligence for Sustainable Development Goals

Abhirup Khanna · Divya Srivastava ·
Ashutosh Mishra · Manoj Kumar ·
Abdul Rajak A. R.
Editors

Artificial Intelligence for Sustainable Development Goals

Principles, Techniques, and Case Studies

Springer

Editors
Abhirup Khanna
University of Petroleum and Energy Studies (UPES)
Dehradun, Uttarakhand, India

Ashutosh Mishra
Birla Institute of Technology and Science Pilani
Dubai, United Arab Emirates

Abdul Rajak A. R.
Birla Institute of Technology and Science Pilani
Dubai, United Arab Emirates

Divya Srivastava
Bennett University
Greater Noida, Uttar Pradesh, India

Manoj Kumar
University of Wollongong in Dubai
Dubai, United Arab Emirates

ISBN 978-3-032-28216-3 ISBN 978-3-032-28217-0 (eBook)
https://doi.org/10.1007/978-3-032-28217-0

This Springer imprint is published by the registered company Springer Nature Switzerland AG
The registered company address is: Gewerbestrasse 11, 6330 Cham, Switzerland

Contents

A Policy Analysis of Artificial Intelligence and Sustainability from a Governance Perspective

Jeffy Johnson and Joel Johnson

Abstract Artificial Intelligence has become a part and parcel of every sector. It is the buzzword in this era of technological advancement. As it is integrated into all spheres, sustainability and governance have many constraints and challenges. The growth of AI enhances innovation, but on the other side, it paves the way for environmental and policy challenges. AI can promote sustainability through global frameworks, which effective use of AI tools to reduce waste, improve energy efficiency, and support sustainable outcomes. The challenges faced are cybersecurity, transparency, human rights, environmental sustainability, etc. Even developing countries like India have taken initiatives such as the National Strategy on Artificial Intelligence, which primarily addresses ethical issues and societal constraints. This is where the role of policymakers becomes vital as they have to ensure perfect equilibrium that fosters sustainability and acceleration of technology. There is a need to revamp the governance framework due to the demands of sustainable development and innovation progress in the ever-changing stage of AI deployment. It is essential to integrate ecological consciousness. The primary objective of this research is to address the sustainability issues in AI governance mechanisms in the frameworks of the United Nations, the European Union, and the World Economic Forum. Further, what methodology should be adopted so that human rights, cybersecurity, and social equity can be devised in the design and application of the AI system through a legal lens. Keeping this in the background to explore the vital ethical issues in AI development and the role of policymaking in the same.

Keywords Artificial Intelligence · Sustainability · Policy · United Nation · European Union

J. Johnson (✉)
Assistant Professor (Law), Bengaluru School of Law and Justice, Bangalore, India

J. Johnson
Master of Computer Applications, Jain (Deemed-to-be University), Bangalore, India

A. Khanna et al. (eds.), *Artificial Intelligence for Sustainable Development Goals*,
https://doi.org/10.1007/978-3-032-28217-0_1

1 Introduction

From being an idea in the futuristic world to being something transformative in so many dimensions, such as healthcare, education, agriculture, and governance, Artificial Intelligence, or AI, has come a long way. Its rapid growth has triggered enormous innovation, giving huge solutions to some of the most problematic challenges facing the world. On the other hand, AI development and deployment are bringing up important ethical, societal, and environmental challenges with increasing concern for responsible governance. Artificial Intelligence has an integral role to play over sustainability as well as international relationships which various countries share among themselves. AI can include various concepts such as neural networks, machine learning and various types of autonomous systems. It can be understood as general-purpose technology, due to its various advantages it is rapidly used across many sectors. But there is always ambiguity in when we examine the policy and governance of the same.

The implications of AI for sustainability also depend on the political dynamics surrounding it. AI can be utilized to protect biodiversity as well as ensuring sustainable development. It has the power to understand and analyze the conditions of environmental problems. Many countries such as USA, China and Russia have showcased willingness to leverage AI to make sure the power balance in their favor in the various ways on factors such as aims, objectives, interest and others. There are many issues and constraints when it comes to policy making process, then it sparks concerns and debates toward decision-making process in the context of AI and sustainability. Only when we delve into the political and policy aspects then alone can we understand and examine the implications of AI for sustainability.

2 Sustainability Equilibrium

The idea of sustainability is heavily dependent upon individuals. The central presumption of the same is conservation of resources and economic development. The economy can develop and grow but the rights of nature have to be saved and not excessively utilized. Artificial Intelligence and sustainability are two fundamental pillars for societal development and growth. With this new era of rapid technology growth AI or Artificial Intelligence is a most critical element in making the world sustainable. The reason why AI can act as a catalyst to sustainable development is that it can scan vast amounts of data, recognize emerging patterns in the system and guide the processes. World Economic Forum, European Union and United Nations act as a control agency so that the AI can be used for good use in a manner to decide sustainable development.

3 Global Frameworks on AI Regulation

3.1 United States

One big step in the journey toward ethical use of AI in the US was made with the introduction of the AI Bill of Rights in October 2022. Backed by the Biden administration, the statement offers a voluntary framework for companies involved in creating, using, and overseeing Artificial Intelligence. Emphasizing transparency and privacy, the AI Bill of Rights aims to ensure that AI systems are fair, inclusive, and transparent and affirm users' rights to understand how their personal information is handled and how they are engaged with by AI systems.

Then next progressive step taken led to the companies should regularly audit their hiring algorithms for discrimination while the New York AI Bias Law takes effect on 5 July 2023. Discrimination that could willfully or unintentionally treat protected classes differently has to be uncovered by means of these tests. In order to encourage greater openness when it comes to applying AI technology to vet potential employees, the law also requires companies to report the outcomes of these probes. As a reaction to the tendency of algorithmic discrimination and AI-driven discriminatory recruitment, the New York AI bias law promotes transparency and responsibility regarding this matter. Upon coming into force in 2023, the New York AI Bias regulation forced companies to keep their checks regularly examine their algorithmic hiring practices for bias. Both deliberate and unintentional bias can cause discrimination. In order to provide transparency in employment of Artificial Intelligence, the companies must publish the significant findings of the audit.

By 2026, data centers in the US are projected to consume 6% of the nation's total electricity. This is because of a spike in demand of AI this leads to a major need for sustainable methods to cater to the needs of AI development and progress.

When AI model are used for the purpose of deep learning it has the capacity to make use of large portions of energy. As per the extensive research conducted by Karen Hao, Senior AI editor at MIT Technology Review, it was discovered that training ChatGPT-3 which is an application which is commonly used worldwide requires energy of 121 houses in the United States for a whole year. The shocking reality was that this process leads to carbon dioxide emissions that is similar to 33 flights between Australia and the UK. Then you may image the energy consumption of GPT-4 will be enormous. It is vital to understand that the use of technology will have a great impact on the environment.

Although state-by-state regulations differ, the majority of current US state privacy laws contain provisions pertaining to the use of automated decision-making and profiling. The majority of states require businesses to offer customers an opt-out for this type of data processing and to make public announcements when they employ AI in automated decision-making. Some places even require businesses to offer customers the rationale for the decisions their AI systems reach and conduct consumer impact analyses.

Due to this greater transparency, customers might be able to gain a better insight into the decision-making process and possibly competition result which appear discriminatory or unfair. These privacy rules are a good start to put companies on their shoulder about how they use AI, and they would be a perfect model for inclusive federal law in the future. But as technology advances so rapidly, these rules will have to be reviewed on an ongoing basis to protect consumers, their environment and stay abreast of advances in technology.

3.2 European Union

Even though the General Data Protection Regulation (GDPR) has its headquarters in the European Union, it focuses on companies from around the globe that handle the personal data of European Union citizens. Specifically, it will ensure responsibility and openness in any company activity dealing with personal information gathering, including recruitment and hiring. Employers are required to make arrangements for GDPR requirements to ensure the right to privacy of EU citizens during recruitment as well as when applying AI. Social Security numbers and fingerprints are just two of the sensitive personal information that your AI solutions must never collect. Secure and appropriate procedures for collecting, processing, storing, transmitting, and deleting candidate data are equally crucial. United Nations Sustainable Development Goals of AI innovations are shifting toward objectives such as responsible consumption, sustainable energy, and climate action, etc.

As part of this regulatory framework, the EU AI Act aims to ensure that AI is applied sustainably and ethically in all sectors of the economy. On March 13, 2024, the European Parliament made the long-awaited AI Act official. Limited-risk AI systems must disclose information, whereas high-risk AI systems are subject to regulations, and low-risk AI systems are frequently left unprotected. Even though the EU AI Act only applies to businesses operating in the EU, vendors that have operations in several countries will likely start to support it, just as what happened when the GDPR privacy regulations were introduced. Such a structure could potentially be adopted by other nations. The only comparable set of regulations for the US at the moment is the Blueprint for an AI Bill of Rights.

According to the AI Act, using AI in the workplace carries a high risk. When choosing an AI solution, hiring managers should consider how it operates and steer clear of those that use biometric information or offer arbitrary sentiment or emotion data. It is also best to stay away from any solutions that eliminate human oversight from the hiring process. For example, deciding whether a candidate should advance to the next stage based only on AI. Employers and recruiters will soon receive comprehensive guidelines that probably address transparency, documentation, and bias reduction. Since many current AI solutions satisfy these requirements, you can seamlessly incorporate them into your tech stack as the law changes and enforcement gets underway.

3.3 India

The AI advice from the Ministry of Electronic & Information Technology (MeitY) takes several proactive measures to address the issues. The latest AI guideline from India stipulates that AI cannot exhibit innate prejudice or discrimination, urges providers to reveal any AI that is not well tested or reliable, and puts safeguards in place to stop deep fakes. Since companies may be held accountable for using solutions that incorporate new prejudices into recruiting or talent management procedures, the prohibition on AI that exhibits inherent bias or discrimination is particularly pertinent to them.

Companies must make sure that human oversight is always present for personnel-related decisions and do thorough due diligence when evaluating AI technologies utilized for talent management in India. On the path to AI-driven sustainability, there are several obstacles in addition to the enormous promise. The challenges that require immediate solutions are numerous and range from algorithmic transparency and cybersecurity dangers to the escalating concerns about environmental sustainability and human rights. Because of the high carbon footprint and power-intensive nature of these models such as neural networks and huge language models eco-friendly development is essential for the industry.

Countries like India have recognized the twin promise and pitfalls of AI. Its National Strategy on Artificial Intelligence, led by NITI Aayog, aims at using AI for inclusive growth while considering ethical issues and societal challenges. The main components of this strategy are as follows:

(a) Developing localized AI Solutions: to help resolve local challenges, including improved agricultural productivity and access to better health services.
(b) Promoting Transparency and Fairness: Ensuring that AI algorithms are free from biases and that operations are conducted transparently.
(c) Encouraging Public-Private Partnerships: Facilitating collaboration between government bodies and private enterprises in an effort to democratize the benefits of AI.

The policy also envisages a robust AI ecosystem emanating from the investment in research and fostering of innovation with its attendant gap areas in digital infrastructure. In effect, that situates the country in global leadership for the ethical and sustainable development of AI.

4 AI Investment by Major Countries over Time (Fig. 1)

The following chart indicates the investment of dominant global players such as the USA, China, the EU, and India in AI research. The numbers indicate exponential growth in the financing of AI over the years with the USA and China taking the lead because of their robust tech ecosystems and governmental support. The EU,

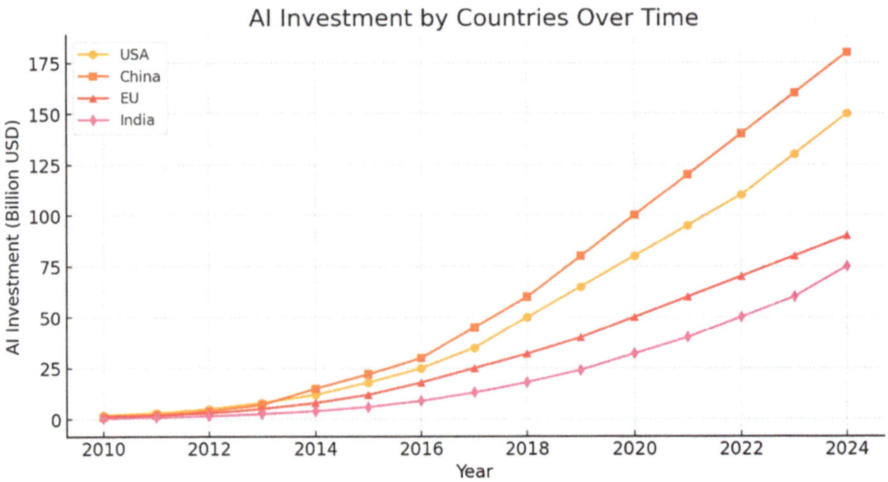

Fig. 1 Data is compiled from these sources Stanford AI Index Report: Global AI Investment Trends, OECD: AI Policy Observatory—AI Investment Statistics, World Economic Forum: AI Global Investment and Funding Analysis

while trailing in finance, emphasizes ethical AI research and regulatory approaches. India's expenditure on AI, although comparatively lower, has been on the rise, with application being higher in the health sector, agriculture, and e-governance. The trend of the graph shows that as AI comes to take a leadership position as an economic and technology output driver, countries are stepping up expenditure so that they are on par at the innovation and automatization level.

5 AI Adoption Rates in Different Sectors in India in FY 2024 (Fig. 2)

The bar chart of AI adoption rates across industries in India for FY24 indicates the variation in AI use across industries. The IT and software industry demonstrates the highest usage of AI, using automation, data analytics, and machine learning for enhanced efficiency. Healthcare is the next one, with AI as a core component in diagnostics, patient management, and drug discovery. The industry is aided by AI-based robotics and predictive maintenance, increasing productivity. AI is being applied on a large scale in finance and banking to prevent fraud, evaluate risks, and automate customer support. AI is also finding increasing acceptance in the retail and e-commerce sector to offer personalized suggestions and streamline supply chains. In general, the graph shows the rate of AI adoption across industries and aiding digital transformation and Indian economic growth.

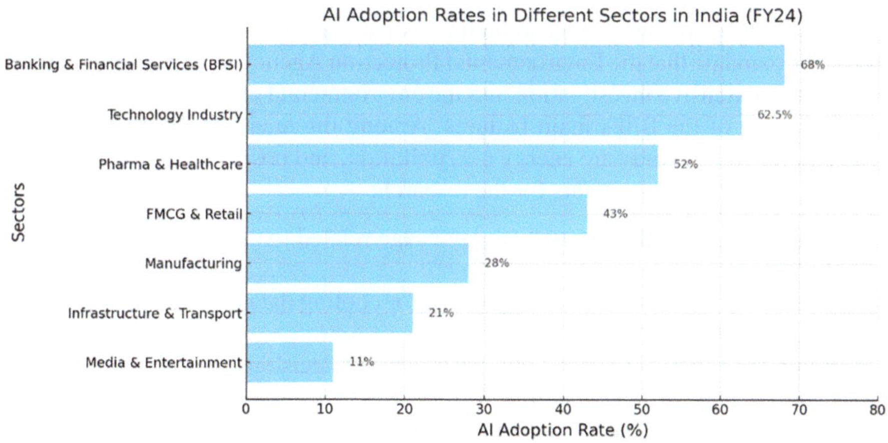

Fig. 2 Data compiled from these sources NASSCOM: State of AI in India, (MeitY), AI Adoption in India—Annual Report, Gartner: AI Adoption Trends in Emerging Markets

6 Artificial Intelligence and Environmental Impacts Act

Sec. 5002. Definitions (3). of the National Artificial Intelligence Initiative Act of 2020, as defined under the Environmental Impacts Act, defines AI as: "a machine-based system that can, for a given set of human-defined objectives, make predictions, recommendations or decisions influencing real or virtual environments. Artificial intelligence systems use machine and human-based inputs to: (a) perceive real and virtual environments; (b) abstract such perceptions into models through analysis in an automated manner; and (c) use model inference to formulate options for information or action." The role of people, the fact that AI is a machine-based system with a variety of outputs, and other characteristics are some of the commonalities between this definition and many others.

Despite being a brief statute, the National Institute of Standards and Technology (NIST) is in charge of the following four important components of the AI Environmental Impacts Act:

1. Researching the effects of AI on the environment and submitting a report to Congress.
2. Forming an alliance of interested parties to research AI's effects on the environment.
3. The implementation of a system that allows organizations to voluntarily report on the effects of Artificial Intelligence on the environment.
4. A study on the environmental implications of Artificial Intelligence to Congress.

All things considered, the Artificial Intelligence Environmental Impacts Act encourages openness, responsibility, and comprehension of AI's environmental impact in order to solve environmental issues related to AI technology. Through voluntary reporting, cooperative research, and standardized assessment, it seeks to guarantee that AI advancements are in line with environmental sustainability objectives. The mandate that the Environmental Protection Agency's (EPA) Administrator carry out an extensive investigation into the environmental effects of Artificial Intelligence is one of the Bill's main features. Among the many implications that will be examined in this study are energy use, pollution, and possible local-scale effects. The following are requirements for the research:

(a) Analyzing the pollutants and energy usage related to AI models and hardware during their whole lives.
(b) Examining the water and energy usage for cooling data centers that are utilized for AI-related tasks.
(c) Examining how hardware kinds, deployment decisions, design, and development affect environmental effects.
(d) Determining possible local environmental effects, such water withdrawals and grid stress.
(e) Evaluating the environmental implications of AI applications, both good (such as increasing energy efficiency) and negative (such as rebound effects).
(f) Resolving divergent negative environmental consequences and taking into account additional environmental impacts as decided by the EPA Administrator.
(g) Including the results of an updated data center analysis that was required by the Energy Independence and Security Act of 2007.

Additionally, during the research, the Administrator will ask for and take into account public feedback.

The Artificial Intelligence Environmental Impacts Consortium is created under Section 5 of the law. The Director of NIST will lead the consortium following consultation EPA, and other pertinent officials. The goal of the collaboration is to develop future standards, metrics, and other factors meant to thoroughly evaluate and record the environmental effect of Artificial Intelligence. University, civil society, and business stakeholders will comprise it.

In addition to encouraging the establishment or development of open-source instruments for measuring such impact, the consortium further intends to ensure there is frequent and consistent reporting regarding the environmental imprint of AI from conception to the point of being discarded and present guidelines on reducing adverse effects as well as maximizing beneficial effects of AI technologies. Understanding and administration of the environmental consequences of AI innovations are the final goals of the consortiums. The Artificial Intelligence Environmental Impacts Act mandates voluntary reporting of the environmental impact of Artificial Intelligence to be mandatory.

The Director of NIST will oversee providing directions to voluntary reporting organizations on how to become a part of the system. These rules will take into account a range of environmental issues, such as energy and water usage, pollution, and e-waste, that are produced in the process the AI hardware and model lifecycle. There will be a comment period before finalizing the rules, and as many voluntary reporting system inputs as are practicable will be submitted.

7 Artificial Intelligence and Public Policy

Public policy application of AI in real-world settings needs an understanding of how stakeholders like the government, academe, private sector practitioners, and consultants build expectations and narratives in a way that it can exert positive social impacts. In describing why, the disparity between visioning of what AI might be able to do in supporting sustainability policy and how it is implemented in such situations is persistent, a sociology of expectations account is used to talk, negotiate, and speculate about what a new technology can or should do.

Collective expectations may structure behavior, give legitimacy and form, stimulate interest, and create investment. Algorithmic and architectural innovation, availability of data, and computing power are all incentives for the evolution of AI technology. But perhaps especially among policymakers, hype about nascent technologies such as AI has a propensity to oversimplify their effect and range of application. Typically, the price of more distant failure is borne by other agents, such as patients, investors, public politicians, the technical and social implementation aspects are usually not emphasized in such representations. Another useful paradigm presented by sustainability transition research will enable us to understand the way AI technology will co-evolve with habits of society, law, practices, and norms to create or fail to create more sustainable futures. The AI development and application for the UN Sustainable Development Goals (SDGs) is based on various factors, of which technical abilities (i.e., model complexity and data availability) and perceived efficacy, social acceptability, proper regulation, and concordance with prevailing policy processes and objectives.

These conclusions regarding transitions to sustainability and sociology of expectations imply that enhanced knowledge regarding the dynamic interdependence of social and technical processes can contribute to better performance in the long term. Examples include publicly acknowledging technology constraints, achieving balance between richness of real-world environments in developing or implementing new technologies, and flexibility and evaluation when applying new technologies to policy environments.

A. Government and Policymaker Perspectives.

As there is not much scientific literature already out on AI government tools or sustainable policy in general, we conducted a search of the gray literature for government AI applications and proposals for AI policy. Sixty countries worldwide have published more substantive AI policies, and dozens more have published or are preparing similar ones. There can be made more provision than is being made, as evinced by a review of national AI strategies and policies in the English-speaking EU, Australia, and the UK. Setting expectations of what can be done with AI to deliver the countries social and economic dividends is the shared goal of any plan or strategy. Although these reports charge Oxford Insights Government Department of Industry, Science, Energy and Resources Government with the responsibility of giving detailed accounts of how AI is or will be used in attempting to attempt to

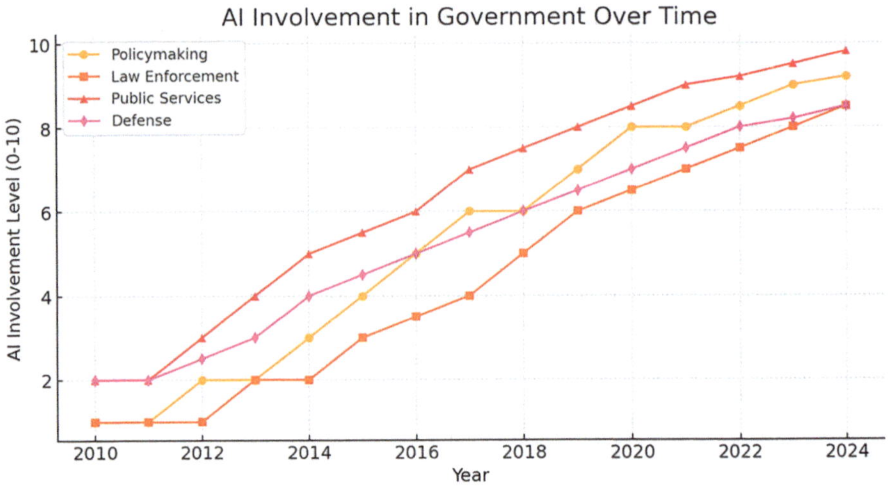

Fig. 3 Data is compiled from these sources McKinsey & Company: AI in government: Capturing the potential value

achieve these objectives, they are all highly interested in the possibility of AI to boost productivity, enhance social welfare, and address environmental issues.

In order to identify patterns in AI development and use by the UK and EU governments, the European Commission hosts the AI-WATCH dataset. Absence of information on the AI tools' efficacy in meeting the purported public need—that is, whether or not they are implemented—defines an analysis of the projects in the dataset. It is true that many of the provided resource links (such as URLs) are no longer operational. As a result, there is a lack of evaluation data regarding the effectiveness of these tools and how they work in real policy contexts. Over Time, AI's Role in Government (Fig. 3).

The AI government involvement timeline graph indicates the way Artificial Intelligence has been introduced into progressively wider public domains such as policymaking, law enforcement, public services, and defense. Governments increasingly rely on AI-based decision-making tools in a bid to sift through vast volumes of data in pursuit of more intelligent policymaking. AI technology like predictive policing and facial recognition has increased in policing, needing more efficiency but also necessitating ethical issues. AI automation has also supported public services in furthering healthcare delivery, education delivery, and the delivery of welfare services. AI has played an increased role in defense as new technologies increased autonomous weapons, cybersecurity, and military strategic thinking. The graph probably indicates a consistent rise, which accounts for the mounting use of AI in government.

B. Role of Policymakers

Policymakers provide the governance frameworks where they provide balanced nurturing of innovation with success in sustainability as required in the future of

AI. The roles of the policymakers include constructing out frameworks in entirety. The governments and international institutions need to join hands in building flexible, inclusive, and enforceable governance frameworks. These must be capable of addressing the rapid pace of technological change while focusing on ethical elements and sustainability goals. Involve various stakeholders Utilize various stakeholders' industry experts, scholars, civil society, and representatives of marginalized communities. This will ensure that ethical and societal issues are dealt with in an integrated manner to foster public trust in AI technologies. Making it accountable the law shall ensure that AI systems are transparent and accountable. For instance, it shall place a requirement on organizations to conduct periodic audits of algorithms that necessitate in line with human rights and environmental sustainability objectives.

C. Advisor and Consultant Perspectives.

The AI government employment solution market is full of players, each with his or her agenda. These players, as well as those who consume AI (e.g., the government) and create it (e.g., data scientists and AI researchers), are necessary. They fill several roles, such as educator, advisor/consultant, knowledge broker, advocate, and technology and institution integrator.

Specialized enablers, such as non-academic analysts and commentators, are employed by think tanks, technology and consultancy corporations, and advisory organizations that are independent or almost sovereign. Their analysis, study, and recommendations for improvement are rarely found in peer-reviewed scholarly literature; instead, they are presented in government proposals, reports, and speeches. Based on publications from consulting companies including the Centre for Public Impact, Deloitte AI Institute for Government, and Deloitte Centre for Government Insights, the benefits and limitations of implementing AI technology at scale are examined and illustrated, along with the role of government adoption. According to their research, evaluating such deployments is challenging because the public sector is still in the early phases of hiring, developing, and deploying AI.

The reminder that there is probably going to be an unequal distribution of sophistication, capability, willingness, or overall readiness for AI tools within governments. "The pace of change in AI capabilities has not been matched by the response of governments," complains Oxford Insights in their yearly cross-country reports on government AI readiness. They demand that the regulatory framework surrounding AI governance be expanded quickly, that public sector capacity be increased, and that they "keep up with global developments and learn what their peers are working on." The report itself does, however, include relatively few examples of real-world government use cases. When these projects are mentioned, links to more information provide a brief description of them and whether they are still active.

8 Challenges of AI Sustainability and Governance

8.1 Cybersecurity

With AI becoming a part of critical infrastructure, it also invites the target of cyber-attacks and data breaches. Strong cybersecurity measures must be in place to ensure the protection of sensitive information and public trust. For example, in the health sector, AI-driven diagnostic tools will need stringent security protocols to protect patients' data.

8.2 Transparency

In this respect, most AI systems are "black boxes," leaving the process underpinning decisions opaque. A lack of transparency erodes accountability and further compounds the challenge of identifying and mitigating biases in algorithms. Consider a scenario with AI-based hiring tools, where there's potential for the machines to discriminate based on biased training data.

8.3 Human Rights

The way in which AI may unconsciously perpetuate discrimination, invade privacy, or curtail freedoms cannot be disregarded. For example, facial recognition systems are found to have a higher rate of misidentification for people from certain demographic groups, opening ethical questions about their use.

8.4 Environmental Impact

Training and deploying AI systems require huge computational power.

To be more specific, one AI model can use energy equivalent to five cars in their lifetime. This gives reason for the urgent development of energy-efficient algorithms and utilization of renewable sources of energy.

8.5 Fairness and Accessibility

The benefits of AI tend to be concentrated in developed countries, so it's increasing the digital divide globally. Poor countries find it difficult to get AI technologies due to a lack of resources, infrastructure, and expertise, which increases socio-economic disparities.

9 Methodology for Embedding Ethical AI Design

The following are important considerations in designing and applying AI systems:

9.1 Human Rights

The biases and discriminatory practices within the AI algorithm should be safeguarded with diverse and representative data sets and mechanisms that can identify and reduce biases.

9.2 Cybersecurity

More solid legal frameworks should address vulnerabilities of AI systems. This will encourage policy makers to set high standards for encryption, routine security audits, and full compliance with data protection regulation.

9.3 Social Equity

The policies should ensure that the gains from AI are well shared and distributed to all socio-economic groups evenly; for example, AI solutions deployed in government programs focusing on underserved areas to reduce the digital divide.

10 Revamping Governance Frameworks for AI Sustainability

It's critical to consider how AI development and use fit into larger sustainability initiatives in order to comprehend the dual nature of AI. How can the development and use of AI technology be guided in a sustainable manner? How can governments and international organizations verify that AI systems help achieve the SDGs? The CSRD takes effect on January 5, 2023, and enhances the requirements for social and environmental reporting. From 2024, additional companies, including non-EU companies, with an annual turnover of over EUR 150 million in the EU will be required to report under the European Sustainability Reporting Standards (ESRS). Various frameworks have been put in place to enable companies to report on their environmental, social, and governance (ESG) challenges. One extremely good example is the Carbon Disclosure Project (CDP), which allows companies to disclose environmental data. Environmental goals, strategic planning, and risk management are some of the data that are utilized. Global Reporting Initiative (GRI) uses a uniform reporting framework for different diverse ESG issues (e.g., human rights, greenhouse gas emissions, labor practices, and community footprint) in an effort to promote global transparency and management processes. Besides advancing the necessity of understanding and integrating AIs unique influence within such ESG reports, requirements of reporting therefore will naturally incorporate AI systems into defining terms although focusing on overall effects on organizations, EU AI Act is another sizeable piece of legislation in Europe operating on the field of sustainability at the time of deployment, training, and developing AI. AI Office and all the Member States in the future will have to define how one needs to stay in touch with as well as minimizing environmental sustainability of AI by harmonized conduct rules and standards. National Institute of Standards and Technology (NIST) approach to the evaluation of AI sustainability also considers model training environmental impact and management practices. By tracking relevant performance measures such as computer resource usage and operating carbon footprint, it seeks to establish measurable sustainability goals and test the trustworthiness of AI systems. In order to minimize the adverse environmental effects of AIs, new regulations and standards place it in question how to quantify and allocate the energy usage that occurs in highly sophisticated, multi-stakeholder AI systems. To meet the demands of an industry and best practices, work processes and reporting frameworks for assessing and quantifying the sustainability risks of AIs are required now.

The governance frameworks for AI need a shift in paradigm toward sustainability and innovation goals. The main areas of focus will be Integrating Ecological Consciousness. The governments and industries should take energy-efficient algorithms and renewable energies for AI operations. For instance, application of green AI principles will be far-reaching in mitigating pollution by AI systems. In the same vein, Google's pursuit of carbon-neutral data centers is an environmental milestone for green AI development. Mitigating Ethical Challenges incorporating ethical principles into AI development ensures that the technology is just, fair, and human rights-friendly. Algorithmic fairness, data privacy, and inclusive access to AI technologies are at the top of policymakers. Fostering Global Cooperation AI has implications that transcend borders and necessitate global cooperation. Aligning national policies with international efforts such as the UNs AI for Good program can produce common problems to be addressed as well as overall progress.

11 Practical Considerations for Sustainability Leaders

In short, it is hard to well track the environmental impact of AI. Strategic assessment processes, a firm allocation of responsibility for sustainability impacts to AI supply chain stakeholders, and consensus on what good mitigations are all required. Some of the measuring issues related to the environmental effects of LLMs are being addressed by open-source tools that are starting to appear in the field of Generative Artificial Intelligence (GenAI).

These methods typically estimate carbon emissions by taking into account four key factors:

(a) The model's training time, which has a direct impact on the quantity of computations required.
(b) GPU energy efficiency, which has an impact on training energy consumption.
(c) Datacenter location, which affects CO_2 emissions per kWh of power utilized.
(d) Possible service provider offsets that might partially offset training's negative environmental effects.

There are still issues with these tools' accuracy, though. For instance:

(a) There is uncertainty over the accuracy and availability of data, particularly in relation to the carbon intensity of data centers and the energy efficiency of GPUs.
(b) Emissions produced during the model fine-tuning or inference stages are not often taken into account because of the excessive emphasis on model training.
(c) Because they have direct access to a model's code or comprehensive technical knowledge, tools are often made for traditional machine learning engineers. Because they could only have access to the foundation models through the application programming interface (API), this could be a limitation for GenAI developers.
(d) Compared to Scope 3 emissions, Scope 1 and 2 emissions receive less attention.

The sustainability consequences of traditional, discriminative AI systems also need to be understood, measured, and reported on, even if LLMs and GenAI in general have recently gained a lot of public attention.

In order to measure the sustainability implications of AI systems in a more comprehensive way, sustainability leaders should take into account the following factors:

(a) Calculate operations: For vendor-hosted models, estimate computing complexity by counting operations or by using the model parameters and tokens.

Using information from cloud providers, calculate power consumption depending on hardware efficiency.

(b) Carbon footprint: Using the data center's carbon intensity or Carbon Usage Effectiveness (CUE), determine the direct carbon emissions while taking into account the difficulties in measuring indirect emissions.
(c) Water usage: Use Water Usage Effectiveness (WUE) on-site and Power Usage Effectiveness (PUE) off-site to evaluate water use.

12 Conclusion

Short of it, it's difficult to monitor the environmental footprint of AI well. There must be strategic evaluation processes, there must be unambiguous delegation of responsibility for sustainability effects to AI supply chain players, and agreement on what good mitigations are. This has led to worries about the environmental footprint and sustainability of AIs, particularly in the wake of the exponentially more sophisticated generative AI models that are being released. The Federal Artificial Intelligence Environmental Impacts Act of 2024 was sponsored by Senator Markey on February 1, 2024, and seconded by a number of other senators. The bill calls for consideration and minimization of the environmental impacts of AI technology as part of the process to hold AI system designers and users accountable and transparent.

It also underscores the need for making the consequences of AIs impact on the environment a reality, and lastly, the Artificial Intelligence Environmental Impacts Act of 2024 is a precautionary approach to balancing the environmental concerns generated by AI technology. To ensure accountability, transparency, and good innovation occur within the AI industry, the Bill would create a consortium, conduct an in-depth study, and implement a voluntary reporting process. Technology playing an ever-increasing role in our future, it becomes important to also look at its wider cultural and environmental factors. Adopting new ways to understand and solve Artificial Intelligences environmental effects would render it more sustainable and focused toward creating a greener and better future. Data access is precious to the UN, EU, and World Economic Forum.

At least some of this is a created perspective, but the majority of this is a product of the way AI algorithms operate. The environmental impact of AI and potential conflicts between algorithmic and other forms of knowledge are two more potential areas of focus that are marginalized. Current ideas of AI are reinforced rather than undergoing a radical change, as evidenced by the frameworks that were found to be comparable to ecological modernization and green governmentality, the institutionalization of discourse, and the emphasis on data.

AI has immense value to companies if utilized in the appropriate manner and with the appropriate safeguards. But the power required to train and release these advanced models, and the enormous quantities of water used to cool the servers, can be devastating to the environment. It will only be fruitful discussion if the advantages of Artificial Intelligence can be potentially being used in order to cater to the constraints created by itself. There needs to be a careful review, identification and evaluation of the public reports. The job done by the AI regulations in various states should be given due credit and there needs to be work done where there are limitations raising. If used effectively and judiciously, Artificial Intelligence tools can offer valuable solutions to sustainability-related policy challenges. Integrating AI into governance and sustainability frameworks is becoming more crucial as it continues to transform the technical and social environments. Finding innovative solutions that will align AI innovation with ethical, social, and environmental goals is the challenge for policymakers.

Bibliography

1. Francisco M, Linner B-O (2023) AI and the governance of sustainable development. An idea analysis of the European Union, the United Nations, and the World Economic Forum. Environ Sci Policy 150:103590. https://doi.org/10.1016/j.envsci.2023.103590
2. Goralski MA, Tan TK (2020) Artificial intelligence and sustainable development. Int J Manag Educ 18:100330. https://doi.org/10.1016/j.ijme.2019.100330
3. World Economic Forum, Center for the Fourth Industrial Revolution (2019) A framework for developing a national artificial intelligence strategy. World Economic Forum, Geneva, pp 4–18
4. World Economic Forum, Bloomberg, Deutsche Energie-Agentur (2021) Harnessing artificial intelligence to accelerate the energy transition. World Economic Forum, Geneva, pp 3–25
5. White House Office of Science and Technology Policy (2022) The blueprint for an AI bill of rights: making automated systems work for the American people. White House Office of Science and Technology Policy, Washington, DC. Available at: https://ai.org.tr/wp-content/uploads/2022/10/9-US-White-House-BLUEPRINT-FOR-AN-AI-BILL-OF-RIGHTS-.pdf
6. Hine E, Floridi L (2023) The blueprint for an AI bill of rights: in search of enaction, at risk of inaction. Minds Mach 33:1–18. https://doi.org/10.1007/s11023-023-09616-w
7. Deloitte (2023) NYC local law 144-21 tackles AI bias in employment decisions: preparing for new rules under New York's legislation. Deloitte Insights. Available at: https://www2.deloitte.com/us/en/pages/audit/articles/nyc-local-law-144-algorithmic-bias.html
8. US Department of State (2024) Artificial intelligence (AI). US Department of State, Washington, DC. Available at: https://www.state.gov/artificial-intelligence/
9. Bolón-Canedo V, Morán-Fernández L, Cancela B, Alonso-Betanzos A (2024) A review of green artificial intelligence: towards a more sustainable future. Artif Intell Rev. https://doi.org/10.1007/s10462-024-10735-2
10. Hao K (2019) Training a single AI model can emit as much carbon as five cars in their lifetimes. MIT Technol Rev, 6 Jun 2019
11. The EU Artificial Intelligence Act (2024) Up-to-date developments and analyses of the EU AI Act. Available at: https://artificialintelligenceact.eu/
12. US Congress (2020) Division E—National Artificial Intelligence Initiative Act of 2020. US Congress, Washington, DC
13. American Institute of Physics (2020) National AI Initiative Act final. Available at: https://www.aip.org/sites/default/files/aipcorp/images/fyi/pdf/national-ai-initiative-act-final.pdf
14. US Congress (2024) S.3732—Artificial Intelligence Environmental Impacts Act of 2024. US Congress, Washington, DC. Available at: https://www.congress.gov/bill/118th-congress/senate-bill/3732/text
15. PwC (2024) What US companies need to know about the EU's CSRD. PwC Insights. Available at: https://www.pwc.com/us/en/services/esg/library/eu-corporate-sustainability-reporting-directive.html
16. O'Connor R, Bolton M, Saeri AK, Chan T, Pearson R (2024) Artificial intelligence and complex sustainability policy problems: translating promise into practice. Policy Des Pract 7(3):308–323. https://doi.org/10.1080/25741292.2024.2348834
17. The White House (2023) FACT SHEET: President Biden issues executive order on safe, secure, and trustworthy artificial intelligence. The White House, Washington, DC, 30 Oct 2023. Available at: https://www.whitehouse.gov/briefing-room/statements-releases/2023/10/30/fact-sheet-president-biden-issuesexecutive-order-on-safe-secure-and-trustworthy-artificial-intelligence/
18. The EU Artificial Intelligence Act (2024) Up-to-date developments and analyses of the EU AI Act. Available at: https://artificialintelligenceact.eu/
19. US Congress (2024) S.3732—Artificial Intelligence Environmental Impacts Act of 2024. Available at: https://www.congress.gov/bill/118th-congress/senate-bill/3732/text
20. Basuchoudhary A, Bang JT (2018) Predicting terrorism with machine learning: lessons from "Predicting terrorism: a machine learning approach". Peace Econ Peace Sci Public Policy 24(4):20180040. https://doi.org/10.1515/peps-2018-0040

21. Oxford Insights (2022) Government AI Readiness Index 2022. Oxford Insights, London, 12 Dec 2022. Available at: https://www.oxfordinsights.com/government-ai-readiness-index-2022

A Review on Artificial Intelligence Enabled Policy and Governance Frameworks for Sustainability

Yashika Dixit, Saumya Goswami, Sarishma Dangi, Abhirup Khanna, Anushree Sah, and Ashutosh Srivastava

Abstract Artificial intelligence is revolutionizing sustainability in numerous sectors like smart agriculture, clean energy, and climate action by using predictive analytics, optimization of resource efficiency, and better decision-making. Although AI has a unique potential to help achieve the UN Sustainable Development Goals, it also brings with it moral, legal, and governance challenges. The paper compares current and relevant policy regimes like EU AI Act, OCED AI Principles, and national United States, China, and India policies and understands the interlock of AI, sustainability, and governance. Those problems that are hardest to solve are algorithmic bias, transparency, data privacy, and the high computation cost of AI systems. The research highlights the need for such governance frameworks of a moral and adaptive nature that can balance innovation with accountability so that AI can be a tool that helps achieve social and environmental goals that are sustainable. AI can be a change agent for transformative order toward sustainable development through the connection of effective regulatory frameworks, global collaboration, and participatory practices. This will help support long-term economic stability and protect the environment. The paper offers a few policy suggestions and research ideas that help identify new risks and better use AI's potential for sustainable development.

Keywords Artificial intelligence · Sustainability · Governance · Ethical AI · Policy frameworks · Sustainable Development Goals (SDGs)

Y. Dixit · S. Goswami · S. Dangi (✉)
TBI-GEU, Graphic Era Deemed to be University, Dehradun, Uttarakhand, India

A. Khanna · A. Sah
School of Computer Science, UPES, Dehradun, India

A. Srivastava
SCSET Bennett University, Greater Noida, India

A. Khanna et al. (eds.), *Artificial Intelligence for Sustainable Development Goals*,
https://doi.org/10.1007/978-3-032-28217-0_2

1 Introduction

AI has transformed every industry through evolving itself as the key character of modern world solutions. However, even though omens for "this future" aim to grant humanity many advantages, they also permit evils to lurk in the shadows. To ensure responsible AI deployment in accordance with Sustainable Development Goals, governments, international organizations, and corporations have implemented policies.

Even though some regulations have been introduced, AI governance is still seen as an ongoing issue, but not a complicated one. It seems clear that these regulations now need to be reviewed and updated. This is a critical moment—governments must promote awareness of human values to ensure that AI can be fully embraced while its risks are effectively managed. AI, Sustainability, and Governance: Relationships will be used throughout our lengthy conversation. What are the real strengths and weaknesses depicted by AI? Their good relationships with innovation and the need for regulation are discussed. Effective adaptive governance can guarantee AI's long-term benefits through initiatives that support social and environmental goals. The chapter's primary focus will be on social issues, the search for the language AI will use, and which of these ideas are most likely to be adopted as laws. AI is a powerful force for sustainability and has achieved remarkable economic resilience over the years, advancing environmental conservation [1].

Our investigation is centered around three fundamental questions.

1. What global and national policy frameworks regulate AI concerning the environment?
2. What are the major factors of the current AI governance model that led to its failure in causing the responsible and sustainable use of AI, such as those that impede the deployment of AI in such a way?
3. In what ways should AI rules be not only morally correct but also flexible to enable AI to have more positive than negative effects on sustainability while the harmful impacts are limited?

This paper is structured in the following way: Section 2 introduces Artificial Intelligence, Environmental Protection, and Sustainable Development Goals (SDGs). Sect. 3 is a literature review focusing on AI's contribution to sustainability. Section 4 is an in-depth study of worldwide policies, and it includes UN AI Ethics Guidelines, OECD AI Principles, and a few national frameworks (EU, U.S., China, India). Sect. 5 is about the problems and discusses possible solutions for future administration. Section 6 wraps up the paper.

2 Background

2.1 *Artificial Intelligence*

AI is one such alternative technology where computers and machines imitate learning, understanding, cognition, decision-making, and creativity. It stands among the excellent components of the computing world today. When given access to large datasets, an AI system typically learns, adapts, and improves, allowing it to identify trends and correlations. It accomplishes this by using algorithms to produce a predetermined set of guidelines or instructions that instruct the AI on how to evaluate information and reach conclusions. A subfield of artificial intelligence called machine learning [2] uses labeled or unlabeled datasets to train algorithms that categorize or predict data. Deep learning [3] is a magnified version of machine learning. It uses artificial neural networks in a multi-layer architecture by which processing data is done. Artificially constructed to imitate the formation and working of the human brain, these networks yield deeper learning and adaptability. Over prolonged exposure to new data, these systems become highly proficient at certain tasks, be they image recognition or language translation, thereby changing the relationship between technology and certain industries in various ways.

AI can be organized in several ways according to the stage of deployment or what actions are performed.

1. **Reactive Machines**: This is the simplest form of AI. It reacts according to stimuli through preprogrammed rules. Such systems cannot store any knowledge; therefore, they cannot change and evolve through experience. Another good example of a reactive machine is the IBM Deep Blue [4] that defeated chess champion Garry Kasparov in 1997.
2. **Limited Memory**: In the present-day AI world, it is said that most of the current ones belong to this group. Unlike reactive machines, most modern AI systems have some sort of capacity to store a little history of past data and thus improve their performance with time. They rely upon training models that "absorb" data so that the systems can detect observations and make sound predictions. Such an AI is called limited memory since, by definition, it is dependent on deep learning as part of machine learning (ML).
3. **Theory of Mind**: This is an AI conceptual stage that has not yet been realized but is actively being researched. This kind of AI would mirror human understanding. It will enable machines to recognize, recall, and respond to human emotions, beliefs, intentions, and social cues. It would mark a huge leap in machine empathetic-ness and social awareness.
4. **Self-aware**: This is the furthest and most speculative form of AI, yet far from existence. This hypothetical machine would be conscious and self-aware. It will be able to make decisions based on awareness of itself and its environment, like humans.

Currently, the more pragmatic manner of classifying types of AI is the functionality of machines. All current AI can be categorized under Artificial Narrow Intelligence (ANI). According to their training and programming, these systems were built to achieve tasks only within a limited scope. For instance, a natural language processing AI algorithm cannot classify objects. Examples of narrow AI are virtual assistants, predictive analytics, and Google Search.

The next level is Artificial General Intelligence (AGI). Such a machine would sense, think, and act like human beings but would not be AGI. The far most advanced level, theoretical, is Artificial Super-intelligence (ASI). It is said that at this level, the machine should surpass human intelligence in everything. Some applications of artificial intelligence technologies that have been much publicized include advanced web search engines (e.g., Google Search), recommendation systems (like those used by YouTube, Amazon, and Netflix), and virtual assistants (e.g., Google Assistant, Siri, and Alexa). Generative tools such as ChatGPT or AI-generated art, as well as autonomous vehicles like Waymo and superhuman strategy-game-playing machines in Chess and Go, are other applications.

Curiously enough, most AI-embedded tools worldwide may now have lost relevancy in terms of being recognized by the common public as AI-made. In the words of one eminent expert, "A lot of cutting-edge AI has filtered into everyday applications, and once something becomes useful and widespread, it often stops being called AI." Fig. 1 is the representation of areas within AI, currently. This demonstrates how AI is gradually becoming a part of modern life, where its very existence is often taken for granted.

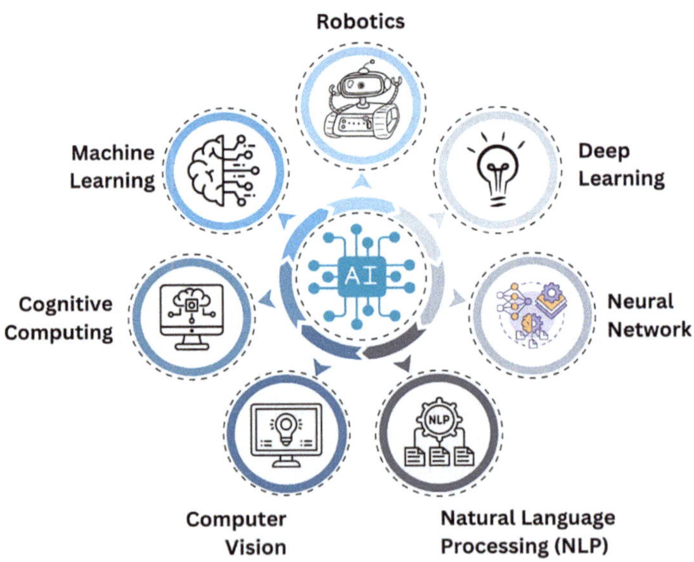

Fig. 1 Key components of AI

At the time AI became an independent structure in 1956, it had undergone several cycles throughout the years, called “AI winters,” during times of great hopefulness followed by relative disappointment and little support. The rebirth came around the year 2012, when deep learning was perceived to surpass other established forms of AI, followed by an enormous rise in investment and research. The movement gained further traction in 2017 with the advent of the transformer architecture, leading to an explosion in AI in the early 2020s. AI comprises numerous research subfields, all concerning different ends such as learning, reasoning, knowledge representation, planning, natural language processing, perception, and robotics. One classic goal of AI is to exhibit general intelligence-the ability for machines to perform any task that is performed by a human being with equal proficiency. Researchers draw from many disciplines in reaching this goal, including mathematics, formal logic, statistics, economics, neuroscience, psychology, linguistics, and philosophy. As of the 2020s, AI and ML emerged as one of the principles of contemporary technology. Applications such as search engines (Google Search), recommendation systems (Netflix, YouTube, Amazon), virtual assistants (Alexa, Siri), self-driving cars, machine translation (Google Translate, Microsoft Translator), facial recognition (Apple Face ID, Google FaceNet), and image labeling rely heavily on it. The use of AI is managed by a Chief Automation Officer in certain organizations. Hundreds of AI solutions specifically targeting various industries are under active use worldwide today.

While AI offers tremendous advantages in speeding scientific discoveries and solving global problems—these also bring severe risks. These include concerns about bias, transparency, and most importantly, ethics, which are relevant to deep learning and other intricate systems. Since AI is already deeply integrated into our daily lives, it is essential to create formal rules and systems to manage its development and use.

2.2 *Sustainable Development Goals (SDGs)*

The 2030 Agenda for Sustainable Development is a plan that was accepted by all United Nations Member States in 2015. It is a worldwide common dream of a tranquil and thriving earth not only for today’s but also for the coming generations. The 17 Sustainable Development Goals (SDGs) constitute the core of this agenda [11]. It is a call to all countries to act urgently on the issue. These goals are intended to eliminate poverty and any other forms of suffering while, at the same time, discovering new ways to develop the economy, improving people’s health and education, and lessen inequality. Also, they involve stopping climate change and the protection of the world’s ecosystems.

The SDGs came about because of the cooperation between the UN and its member states over several decades. Their basis was the 1992 Rio de Janeiro Earth Summit [5]. At this forum, more than 178 nations committed to Agenda 21, a detailed plan for global collaboration in sustainable development. The Johannesburg Declaration on Sustainable Development in 2002, apart from that, went on to stress poverty eradication and environment protection as it was stated in Agenda 21 and the Millennium

Declaration by supporting the implementation of partnerships among all the different agents.

The 2012 Rio + 20 United Nations Conference [6] on Sustainable Development marked a significant breakthrough. During this summit, member states adopted the outcome document to the establishment of the UN High-level Political Forum on Sustainable Development titled "The Future We Want," which initiated the process of developing the SDGs and led. The General Assembly created an Open Working Group of 30 members in 2013 for drafting the Sustainable Development Goals. In January 2015, governments started the intergovernmental negotiations on the post-2015 developmental agenda. The process of the United Nations Sustainable Development Summit held in September, and the adoption of the 2030 Agenda for Sustainable Development took place there, was the final mark of the journey.

2015 was the year when the global community had to cooperate closely to overcome challenges, and several significant agreements were made because of this cooperation. Individually, they consisted of the following: the Sendai Framework for Disaster Risk Reduction (March), the Addis Ababa Action Agenda on Financing for Development (July), the 2030 Agenda (September), and the Paris Agreement on Climate Change (December).

The High-level Political Forum is the main organ of the United Nations, which is responsible for the review and oversight of the implementation of the SDGs. The division for Sustainable Development Goals (DSDG) in the UN Department of Economic and Social Affairs (UNDESA) is committed to achieving the goals and provides the necessary resources and capacity building. It takes care of issues such as water, energy, climate, oceans, urbanization, transportation, science and technology, and partnerships, for instance, that of the Small Island Developing States. Besides that, the DSDG is leading the way for the UN to implement the agenda, producing the Global Sustainable Development Report (GSDR), as well as participating in and facilitating communication and advocacy activities. Global ownership on a wide scale and strong commitment of all stakeholders to implement the SDGs are the conditions without which the 2030s will not succeed.

2.3 Sustainability and Artificial Intelligence

The Industrial Revolutions were the major events that introduced into the world a culture of innovation. This eventually led to the inventions of new technologies such as artificial intelligence (AI). AI is one of the technologies which can tremendously speed up the implementation of the United Nations' Sustainable Development Goals (SDGs) that are the global economic, social, and environmental sustainability development framework [1]. Besides technical issues, AI is also regulated by various laws that address concerns such as liability, privacy, non-discrimination, and product safety. Regulatory initiatives specifically targeting AI are becoming increasingly numerous as well. The formal going into effect of the European Union's Artificial Intelligence Act (EU AI Act) in 2024 was a major step forward. Like the General

Data Protection Regulation's (GDPR) expected worldwide influence, the EU AI Act is expected to influence the regulatory strategies of many jurisdictions.

Legislation talking about AI has been more than twice as many in 2023 than in 2022, which means that policymakers worldwide have put AI high on their agenda. One regulatory measure that mainly deals with generative AI is China's Interim Administrative Measures for Generative Artificial Intelligence Services. Apart from the United Nations Educational, Scientific, and Cultural Organization (UNESCO), the US National Institute of Standards and Technology (NIST), the Organization for Economic Co-operation and Development (OECD), the International Organization for Standardization (ISO), and the Group of Seven (G7) have been the major leaders of the initiatives that have resulted in increased worldwide cooperation and the setting of standards concerning artificial intelligence.

The expansion of safety-related projects in the US, UK, Singapore, and Japan as well as the founding of new AI safety institutes is some of the international community's recognitions of the increasing concern of AI safety that took place in 2024. The establishment of the EU AI Office under the EU AI Act is another example of the region's commitment to developing and promoting best practices for AI governance. As the regulatory environment around AI is getting more complicated, the significance of international agreements on interoperable standards and basic regulatory requirements is becoming more visible. These documents will be the main instruments in facilitating innovation, at the same time, ensuring safety and ethical alignment. In this respect, numerous organizations are taking the initiative to implement self-governance models to be in harmony with their internal values, as well as to generate trust and obtain credibility. The start of such activities frequently means that the organization not only complies with legal requirements but also exceeds them by adhering to higher ethical standards. Voluntary frameworks such as the US NIST AI Risk Management Framework, Singapore's AI Verify framework, and UK AI Safety Institute's open-source Inspect platform are good examples of instruments that support organizations in implementing responsible AI practices.

Both organizational and technical controls are necessary for the efficient governance of AI systems. The Ethical AI Implementation hierarchical framework illustrated in Fig. 2 is an example of how organizational management can assist in implementing ethical principles in AI which is evident in the ISO/IEC 42001 [7] international standard. Foundational mechanisms for oversight are provided by organizational management structures, such as those described in the ISO/IEC 42001 international standard. At the same time, technical controls integrated into socio-technical systems play a vital role. Automation is progressively backing up these measures by means of functionalities like AI red teaming (a systematic way of checking AI models for weaknesses), metadata recognition, logging, monitoring, and real-time alert systems. Automated governance tools will be indispensable to the continuation of AI developments as they will become faster and more powerful. Harmonized standards and technical safety controls are expected to gain widespread adoption as they evolve. However, human-AI collaboration will still be a very important part of responsible AI governance. An excellent illustration of how businesses can use both organizational and technical governance solutions at the same time is IBM's

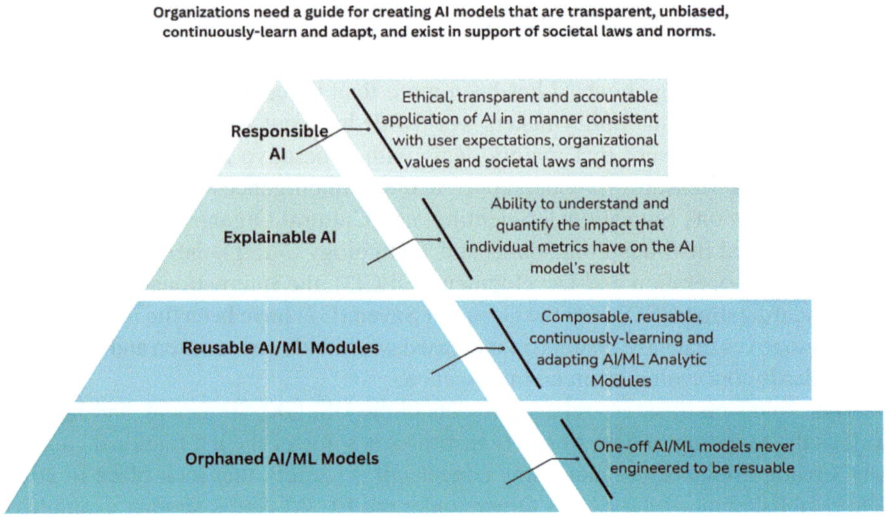

Fig. 2 Ethical AI application pyramid

creation of an AI Ethics Board and Governance Program in conjunction with its Watsonx.governance tools [8].

2.4 *Contributions of AI to SDGs*

One of the most significant AI enablers is the United Nations Sustainable Development Goals (SDGs) that are a source of major solutions to some of the most pressing problems on Earth. In fact, AI dominates most of the industries to make better decisions, increase productivity, and keep the promise of using resources as efficiently as possible to speed up the 2030 Agenda for Sustainable Development.

Precision agriculture under the Zero Hunger (SDG 2) is radically changing food production through monitoring the health of the crops, the type of the soil, and the atmosphere using satellite images, Internet of Things devices, and machine learning. Data-driven choices, such as cutting pests, fertilizing, and watering, are given to farmers by these technologies thus the yields are skyrocketing while Mother Nature is maintained. Besides that, the implementation of AI-based supply chain optimization platforms through improving inventory management and distribution logistics is cutting food waste by about 30%. Thus, sustainable food systems are being realized [9].

The AI technology has ushered the healthcare sector (SDG 3) into a new era with fancy and outstanding accomplishments, especially in the diagnosis [10] and treatment areas. For instance, physicians are empowered with deep learning algorithms to diagnose different diseases from imaging data with even more accuracy than that of human experts. This has led to better patient outcomes and early interventions. Public health officials can observe the spread of diseases, and thus, they can take rapid actions thanks to predictive analytics; on the other hand, the AI telemedicine platforms are facilitating access to quality healthcare in the remote areas. To demonstrate its full potential in global health emergencies and to allocate resources efficiently, the AI model played a crucial role in the tracking of virus spread during the COVID-19 pandemic.

Artificial intelligence is dramatically changing education systems (SDG 4) using personalized learning platforms that automatically adjust to every student's distinct requirement and learning styles [11]. With the help of natural language processing, instructional material can be translated instantly which facilitates the overcoming of language barriers thus more people have access to quality education. AI-powered tutoring programs provide students with on-demand assistance and correction while auto-grading software teachers' time for highly valuable face-to-face student interactions. All these advances are of tremendous help in the educationally disadvantaged areas.

The transition to renewable energy is being facilitated by smart grid management systems in the energy (SDG 7) sector that rely on AI to match supply and demand. With the help of machine learning algorithms that optimize the output of wind turbines and solar panels, the efficiency is raised by 20%. While intelligent building management systems monitor the usage patterns to facilitate energy reduction in the cities through automation, the AI-enabled predictive maintenance is responsible for the minimal downtime of energy infrastructure.

AI is helping the Industrial Sector (SDG 9) to meet its challenge of lessening environmental pollution and wastage due to manufacturing processes. Predictive analytics lighten the supply chains and inventory management tasks, and computer vision systems are much more efficient than human inspectors in spotting defects in the production lines at the early stages. Furthermore, the very same artificial intelligence (AI) systems also power smart city projects that upgrade public services, relieve traffic, and enhance transport systems, all of which make a city more livable and sustainable.

Some AI implementations for Sustainable Cities and Communities (SDG 11) entail integrated waste management systems that employ computer vision to recycle more intelligent traffic management that not only optimizes traffic but also cuts emission. In addition, AI can be very useful in disaster recovery and the location of disaster intervention.

With the help of very refined models, which forecast both climate and extreme weather events, artificial intelligence is an important factor for climate action (SDG 13). Machine learning algorithms are utilized to make the carbon capture operation more efficient, and at the same time assist the enterprises to keep track of, and even lower, their carbon footprints. Besides that, AI-based monitoring systems can be on the lookout for environmental crimes and illegal logging.

AI conservation measures for marine and terrestrial ecosystems (SDGs 14 and 15) include the use of species monitoring that relies on the interpretation of acoustic and visual data to track animal populations. Besides monitoring forests with the help of satellite data and machine learning, these technologies also monitor illegal fishing. By being aware of an ecosystem's health, conservationists can be more effective in protecting it.

The designing of AI for sustainable development should be in line with the implementation of all 17 SDGs to bring about a more sustainable and fairer global environment that benefits everyone, notwithstanding issues like ethics, data privacy, and the creation of digital infrastructure for the less privileged countries. This can be achieved with the right governance and mutual support between countries.

3 Literature Review

Artificial intelligence (AI) sustainable multidisciplinary literature has a breadth of different disciplines [12], focusing on diverse environmental themes, of which are the most represented and transportation, energy systems, water management, and biodiversity. ML models are the leading force, and they are primarily focused on investigating environmental, economic, and to some extent, social aspects, with the most significant attention being paid to climate change and smart cities. For example, in water networks, ML is used for water quality assessment and streamflow prediction, and the research is going on for such tools like linear regression, genetic algorithms (GA), artificial neural networks (ANNs), and adaptive neuro-fuzzy inference systems (ANFIS). In biodiversity, ML and natural language processing (NLP) are used in studying ecosystem services, social sentiment on social media, and habitat suitability. The leading-edge AI-energy applications, such as ANNs, expert systems, pattern recognition, and fuzzy logic, are directed toward building energy consumption, renewable energy, system operations, and public perception [13]. Most of the present work is situated mainly in the technical and engineering fields, where the focus is more on problem-solving than on theoretical intersections between AI and sustainability, thus creating a lot of unexploited potential for disciplines like Information Systems (ISs). The current review is an outcome of the research done on the Web of Science and Google Scholar (2015–2019) and AI-sustainability research saw a rapid expansion during that time, with many of its seminal ideas being derived from the earlier works [14]. Apart from data-driven learning, AI is also about human thinking emulation, for example, Computer Vision (CV) and NLP. CV deals with visual information, while NLP offers machines the ability to understand and generate human language. Semantic computing is a combination of NLP and software engineering that aims to understand user intent.

Similarly, to human beings, with the help of AI technology, machines can have qualities like affective computing (emotion recognition), fuzzy logic (uncertainty reasoning), and expert systems (knowledge-based problem-solving). Progressively, increased ML and fuzzy logic are utilized in expert systems, which is better decision-making. AI robotics, through combining cognitive, emotional, and reasoning skills, is a robot capable of performing human-like activities in the fields of manufacturing and healthcare [15]. However, AI robotics is still far away from achieving full human-like intelligence (AGI), and it already is a very good strategist and can learn on its own.

AI is one of the essential tools in reviving smart cities through the provision of more efficient transport and resource management. Smart cities are the ones that ensure livability and efficiency through agile governance [16] and the six dimensions of smart cities being economy, mobility, environment, people, living, and governance. Services, governance, innovations, culture, and sustainability are the outcomes of an advanced IT infrastructure coupled with AI-driven analytics of the big data that IoT is generating [17]. Knowledge-based AI systems also play a supporting role, while ML, usually combined with CV and NLP, is the main driving force of this analysis.

4 Policy and Governance Frameworks

As artificial intelligence increasingly enters human life, global lawmakers are working to create governance models. Not surprisingly, countries are advancing varied AI governance strategies based on their specific legal, cultural, and institutional contexts. For example, on May 11th, the EU's Artificial Intelligence Act was voted on by the European Parliament, seeking to prohibit or restrict high-risk AI. At the same time, the UK is seeking "regulatory innovation" to become an "AI superpower," as set out in Michelle Donelan's white paper, and has in training the pro-innovation Artificial Intelligence and Data Act (AIDA). Canada too is developing its AI regulation framework with Bill C-27, including AIDA. Singapore takes a balanced approach, as demonstrated by its Model AI Governance Framework. Conversely, China's generative AI draft measures give precedence to social values, ethics, and content standards.

4.1 United Nations AI Ethics Guidelines

The Principles for the Ethical Use of Artificial Intelligence throughout the UN system were formally approved by the Chief Executives Board for Coordination of the UN System in September 2022. The High-Level Committee on Programs (HLCP) provided direction for the development of these principles, which were adopted at an intersessional meeting in July 2022. The development of these principles was a collaborative effort led by the United Nations Educational, Scientific, and Cultural Organization (UNESCO) and the Office of Information and Communications Technology (OICT) of the UN Secretariat. Both entities co-chaired the dedicated workstream within the HLCP Inter-Agency Working Group on Artificial Intelligence. These principles (listed in Table 1) are rooted in the Recommendation on the Ethics of Artificial Intelligence, which was adopted by UNESCO's General Conference during its 41st session in November 2021. Ten fundamental ethical and human rights-based principles make up this framework, which is meant to direct the responsible application of AI in all stages of the system lifecycle-from design and development to deployment and monitoring-across all UN system entities [18].

4.2 OECD AI Principles

The member countries of the Organization for Economic Co-operation and Development (OECD) combine their efforts to find solutions to complex global problems in the social, economic, and environmental areas. Besides being a forum for exchanging policy experiences, the organization also acts as a platform for benchmarking and coordinating national and international policies. Moreover, it is a major player in the development of policies in the new areas of corporate governance, the digital economy, and demographic changes caused by aging. The EU, with which the OECD has a special observer status, is one of the 38 members of the OECD including the US, the UK, Germany, Japan, and Australia in 2024.

In 2019, the OECD AI Principles was the first globally recognized set of intergovernmental standards for the development and use of AI in a responsible way. They were updated in 2024 to a large extent to AI systems which are reliable, creative, and respect basic human rights and democratic principles. The so-called framework consists of five values-based principles and five practical recommendations thereby forming a moral guide for stakeholders and policymakers of the AI ecosystem. Governments and organizations can direct AI development in a way that creates the greatest societal benefits with the least possible risks if they follow these principles. Such a platform would generate the social use of AI as a means of social justice, sustainability, and human dignity. According to the OECD.AI Policy Observatory, as of May 2023, more than seventy different jurisdictions had introduced over 1000 AI-related policy initiatives, thus showing a worldwide commitment to the regulation of AI in accordance with the OECD AI Principles [19].

4.2.1 Value-Based Principles by OECD

The goals of the OECD framework of principles are based on ethical values for AI development and use. The focus of this is to do as much good as possible for society. Firstly, for this to happen sustainably, with inclusive growth and the well-being of everyone being the overall result, the different stakeholders must work together and at the same time Trustworthy AI must be employed. The application of AI should be aimed at human potential, motivating innovation, engaging disenfranchised communities, and equalizing gender, social, and economic disparities. AI-driven systems, in return, have to be compatible with the UN Sustainable Development Goals set for human welfare as well as for the natural world.

One more very significant factor at this point is human rights and democratic values. The law must be respected, and AI solutions must adhere to social justice,

Table 1 UN ethical principles for responsible AI development and deployment

Principles	Description
Do not harm	AI systems should never cause harm whether to people, society, the economy, or the environment. Every phase of AI-from design to deployment-must respect human rights and freedoms. potential impacts must be closely monitored and minimized
Defined purpose, necessity and proportionality	AI should only be used for clearly defined, justified purposes that align with the UN's mission. The method used should be appropriate and not excessive for the intended outcome
Safety and security	Risks to human safety and the environment must be identified and managed across the AI lifecycle. To guarantee that the AI system is safe and secure to use, robust frameworks should be in place
Fairness and non-discrimination	All people must be treated equally by AI systems, free from prejudice, discrimination, and unjust human rights restrictions. Everyone should fairly share the risks and rewards
Sustainability	AI ought to support social, economic, and environmental sustainability. Negative effects must be minimized, and its influence on current and future generations should be regularly evaluated
Right to privacy, data protection and data governance	Privacy rights and personal information must always be respected. to guarantee the integrity and appropriate use of data, robust governance and data protection systems are necessary
Human autonomy and oversight	AI should never take precedence over human autonomy or judgment. Human oversight is essential, particularly when making decisions involving rights or life or death. Human control must be at the center of AI design
Transparency and explainability	Decisions made by AI must be comprehensible and explicable. People should understand the reasoning behind decisions made by AI that affect their rights or benefits
Responsibility and accountability	There must be well-defined procedures for supervision and accountability. There must be procedures in place for both investigation and repair if harm is caused. people, not just the system, need to be held responsible
Inclusion and participation	The development of AI ought to be inclusive and collaborative. The use of AI should be shaped by all stakeholders, including under-represented groups. Diverse perspectives and gender equality are essential

equal treatment, freedom, human dignity, privacy, and data protection principles. Also, it is expected that AI entities will be at the forefront of the fight against misinformation and disinformation in general and that of AI-generated especially. In doing so, they should be aware of the need to weigh freedom of speech against other internationally recognized rights. Moreover, according to avant-garde technical criteria, the establishment of steps and supervisory authorities in this respect is envisaged to apprehend and control any unexpected incidents, abuse, or uses that go beyond their original domain. One of the most important elements for establishing that trust is explainability and transparency, which complement the other essential topics. To enlighten the public on the working methods of AI systems, AI entities must appropriately disclose and communicate the relevant and understandable pieces of information. They should highlight the capabilities as well as the limitations, and if users are to be informed about their interactions with AI, especially in the professional realm, then the data sources, reasoning, and decision-making should also be explained.

To guarantee the stability of such solutions from the beginning up to their implementation and even later, security, safety, and robustness must be supported. If the undesirable and hazardous situation of AI systems comes to be, then there should be the possibility of overriding, interfering with, or stopping the concerned AI units in a secure manner. To guarantee the integrity of information and the observance of other important rights, like that of free speech, such steps should also be taken in all the places where it is technically possible.

Finally, accountability is a very important idea which regulates the conduct of all AI players and is one of the keystone concepts in the framework of AI ethics. The promotion of this mentality will certainly involve the retrievability of datasets, decision-making processes, and system output at every phase of the AI lifecycle. Depending on each actor's individual function and capability to influence the system, criteria for a structured and ongoing risk management approach should be laid out. To prevent the arising of bias, violation of human rights, risk to safety, breach of privacy, as well as labor and intellectual property rights, developers, suppliers, users, and other stakeholders should perform their respective functions cooperatively.

4.2.2 Recommendations for Policy Makers Incorporated by OECD

The OECD put forward several practical proposals to harness AI's potential in an ethical, lawful, and social way [20]. The machine model guidelines should direct national policies that have long-term plans for inclusive international collaboration in fair AI research and implementation. A government-driven open environment, where not only technological progress but also the ethical, legal, and social aspects of AI are funded through interdisciplinary approaches, requires a substantial investment by the public sector. Besides that, they should make it possible for open, high-quality, and representative datasets to be created, with due regard to data privacy, in order to lessen bias and to enhance interoperability through standardization. Finally, governments should develop an inclusive AI ecosystem by creating interoperable digital infrastructures (open data, AI tools, computing) with robust connectivity and by promoting innovative mechanisms such as data trusts for ethical and secure data sharing, particularly in last-mile AI development. It is equally essential for governments to put in place a robust policy framework that facilitates easy transitions from the research stage to the practical application stage. An important element in this regard is the decision to operate regulatory sandboxes aimed at experimentation and the gradual, responsible deployment of AI systems, thus giving preference to outcome-based strategies rather than merely specifying the measures. Lastly, international coordination is necessary to achieve legal interoperability and coherence in AI governance, necessitating ongoing policy and legislative evolution to further innovation, competition, and trustworthiness. Policymakers should react to the consequences of AI in the employment sector by upgrading human abilities via lifelong education, vocational training, and digital skills. State intervention should support welfare measures for the displaced workers and also contribute to employment generation with the equitable distribution of benefits from AI. There cannot be trusted AI without international collaboration. Governments, particularly in the developing world, must participate actively in international bodies such as the OECD to exchange knowledge and create long-term collaborative capabilities. Cross-sectoral and multi-stakeholder efforts must be created to form consensus-based technical standards on interoperable AI systems. Internationally comparable indicators for AI development and use are also needed to track research, development, and deployment, track policy effects, and enhance responsible AI stewardship. Figure 3 emphasizes the interconnection of societal, individual, and technological elements in creating ethical and Trustworthy AI systems.

4.3 European Union AI Regulations

The Artificial Intelligence Act is the first comprehensive piece of legislation of the European Union concerning artificial intelligence. It is a historic regulation made by the European Commission that was made public on April 21, 2021. The new regime is founded on a risk stratification classification system for AI applications which is intended to guarantee that AI technologies employed in the EU are climate-friendly, do not discriminate, are safe, transparent, and traceable. One of the Act's primary clauses mandates ongoing human supervision of all AI systems to prevent any unfavorable outcomes. The European Parliament has also emphasized the need for a very flexible and technologically agnostic definition of artificial intelligence to be adopted, given the dynamic nature of AI development. With layered compliance requirements for developers and users, the AI Act's regulatory architecture groups in AI systems according to the risks they might pose. The so-called unacceptable risk of AI applications is those that are most strictly prohibited. These include systems that: (1) affect the cognitive behavior of individuals or vulnerable groups (e.g., voice-activated toys encouraging dangerous behaviors in children); (2) encourage social scoring based on

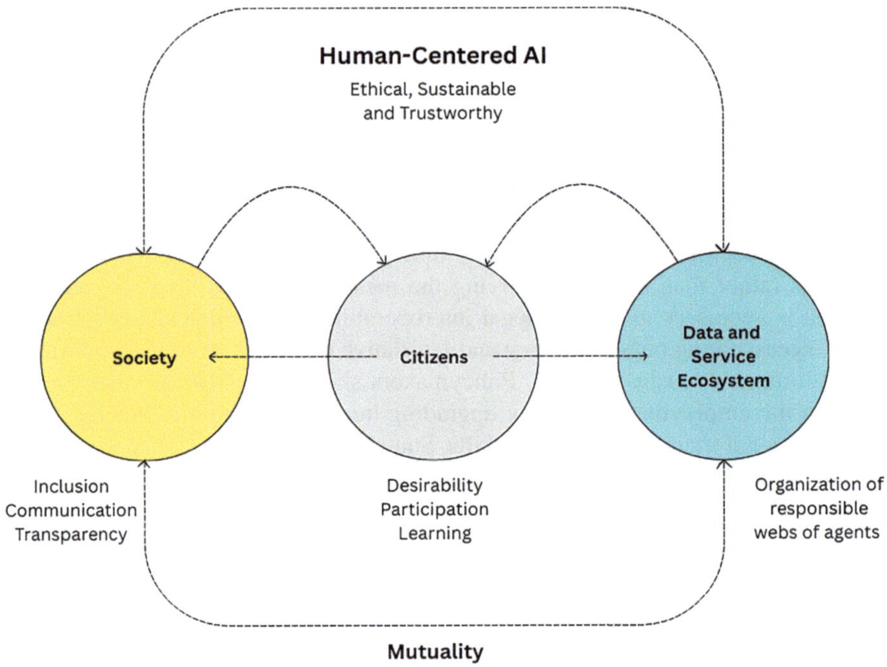

Fig. 3 Conceptual framework for human centered AI: balancing society, citizens, and data ecosystems with ethical principles

behavioral or socioeconomic characteristics; (3) use biometric data for identification and classification; and (4) permit real-time remote biometric surveillance in areas that are open to the public. There are very strict ex-ante judicial controls over the limited exemptions that apply to applications related to law enforcement, particularly regarding post hoc identification protocols [21].

High-risk AI means in the first-place systems which might endanger safety or breach basic rights. As a result, the integrated AI component(s) in presently regulated products (e.g., medical devices, automotive safety systems, aviation equipment, and elevator/mechanical escalator industry) fall under this high-risk category. Besides this, the EU hosts a register on which any AI implementation in the area that is sector-wise (energy, water supply, health sector, transport, telecommunication, and education) through a listed use case e.g., school assessment, employment decision, and so on, must be declared. Before such devices can be offered on the market, they must undergo an extensive compliance assessment. Their perpetual reviews during operation are equally necessary. Besides that, it lays down the "system" or formal routes through which concerned persons may whistle-blow on these "systems" to the national supervisory authorities whose names are indicated in the document.

Even if ChatGPT and other generative AI models fall under the non-high-risk category, they still must abide by some transparency rules. According to these rules, AI-generated content must be identified, architectural measures must be in place to prevent the manufacture of illegal content, and the training material must be summarized and made available with particular attention to copyrighted material. General-purpose AI models with system-wide risk, such as GPT-4, must also go through comprehensive risk evaluations in connection with their obligation to notify the European Commission of significant incidents. Moreover, any artificially generated or altered media content like synthetic voice, images, or videos should be readily identifiable so that users can be informed.

The AI Directive has segments that enable AI to become a facilitator for research and development within the EU. Member states, in an effort to create regulatory sandboxes for start-ups and small and medium-sized enterprises (SMEs) to experiment with AI systems pre-market, need to establish masked operational conditions for controlled testing. That is the way Europe supports its AI innovators to be at par with global competition. A different group in the European Parliament along with the newly formed AI Office in the EU Commission is handling the implementation of the AI act. This collaborative governance structure is designed to bring regulatory clarity, compliance support, and a firm basis for the growth of Europe's digital economy. The regulation which was officially adopted in June 2024 charts stages for conformity: The implementation of operational codes of practice is expected nine months after the execution of prohibitions on unacceptable risk systems, there is a twelve-month period before transparency requirements for general-purpose AI come into force, and high-risk systems are to be fully compliant within 36 months from the law's entry into force.

4.4 U.S. AI Policy and Sustainability Initiatives

The U.S. governance system of the AI ecosystem is based on the principles of executive, legislative, and regulatory actions that focus on the responsible development and application of AI technologies [22]. At the executive level, President Joe Biden's October 2023 Executive Order 14110, "Safe, Secure, and Trustworthy Development and Use of Artificial Intelligence," defines policy priorities for AI innovation and competition, consumer privacy and civil rights protection, and retention of technological leadership in the United States. The issuance of the order has set the institutional basis for AI governance as federal agencies are required to comply with the order by appointing Chief AI Officers.

Additionally, several legislative measures have been added to this executive action to manage the coordination of federal AI research and development in a way that would benefit both national security and economic interests. One such measure is the National AI Initiative Act of 2020. The proposed laws such as the Algorithmic Accountability Act would make companies responsible for identifying the sources of bias in automated decision-making systems and applying the necessary methods to prevent such bias. Different regulatory authorities have also over time come up with their own norms; the FDA has set a direction for AI applications in medical devices, which mainly stresses the areas of accountability and transparency, whereas NIST has delivered AI Risk Management Framework. Besides, the United States has been perfecting its AI governance standards to be compatible with those of other countries via the OECD and Global Partnership on AI. These concerted domestic and international regulations aim at laying down a robust AI governance framework capable of resolving moral issues, safeguarding basic rights, dealing with anxieties about national security, and facilitating technological progress as well. The variety of strategies used by the United States shows its efforts to rank among the leading economies by handling these extremely complicated issues resulting from the diffusion of AI in a tiered governance structure.

4.5 China's AI Governance Model

Compared to Western models, this model is much more centralized and allocates state funds and resources to AI applications rather than addressing corporate social responsibility or, in fact, individual rights in relation to AI. The foundation of the AI strategy was laid out in the State's 2017 New Generation Artificial Intelligence Development Plan, which set ambitious goals to make the country a global leader in AI by 2030. This state-led initiative is designed to improve innovation through investment in research while keeping strong government supervision to maintain stability in the political and economic environment. China's set of laws governing the use of AI at present highlights data governance and the management of algorithms as the main aspects. Unlike the EU's GDPR, China's Personal Information Protection Law provides that state authorities have more extensive access to data while still maintaining some provisions for individuals' privacy. Besides, there are some other rules such as the Algorithmic Recommendation Guidelines, which also regulate the use of algorithms, i.e., require that automated decision-making systems adhere to state censorship policies and, at the same time, disclose, in particular, on digital media platforms, that they are giving the user a certain option by the algorithm.

The most significant aspect of Chinese regulation, perhaps, is the use of AI for internal security. To make a city smart, among other things, upgraded monitoring measures with facial recognition and predictive analytics have been implemented and have consequently become the leading means of law enforcement, which, in turn, is primarily empowered by these technologies. Nevertheless, the method adopted contradicts sharply with worldwide public worries over privacy assurance and control. The examples of policies' correspondence reveal how they find a balance between political power and technological progress. The government has set the rules for generative games very tightly, which, in effect, means that big tech companies are obliged to use the filtering systems developed by the state to prevent the way of information which has not been approved from spreading. This is done in the hope that it will still align with the party's own narrative.

A comparison of the style of ruling of China with that of the West reveals that their priorities and values of each are vastly different. For example, China is very much determined to practice good governance, respect laws, and keep order in the society, apart from that, the state security is also a priority. However, Europe and the United States, on the other hand, are more likely to focus on individual rights and corporate social responsibility. Besides, the difference is reflected in different standards and laws being followed; i.e., in China, majority of the regulations are derived from government policies while in the West there is more focus on democratic accountability and impartiality. The well-planned, government-supported approach helped China not only to quickly upgrade its technological capabilities but also to become a dominant power in the world. In fact, sectors in China have invested more than $26 billion (about $80 per person in the US) in AI-related ventures in just 2023. Guardhouses' perceptions of their new strategies for governing AI in developing economies are being shaped by imported surveillance technologies and smart city infrastructure, which offer an alternative to Western regulatory paradigms [23].

In this way, society can keep political stability and at the same time specify AI governance as a part of national development goals. Certainly, there are some trade-offs related to this model in terms of the individual's rights to freedom of speech and privacy. China's architecture continues to be a potent point of reference as global AI governance develops, influencing traditional Western conceptions of technology and democratic ideals. China's legal frameworks are summarized in Table 2.

4.6 India's AI and Sustainability Policies

India is leveraging AI to solve environmental problems in the above-mentioned sectors. The use of AI in such areas is in line with the country's overall goals of advancing science and technology while abiding the sets of SDGs. The government, academic community, and private sector have been collaborating intensively in opening up new avenues for AI to impact governance, management, and field work [24]. By policy, NITI Aayog's 2018 National Strategy for Artificial Intelligence devoted five main sectors—healthcare, agriculture, education, smart cities, and mobility—a significant push to lead AI adoption in these areas. The direct successor to this framework is RAISE 2020, which has the objective of facilitating better coordination among the stakeholders to create AI applications that are more inclusive and environmentally friendly. MeitY took a step further in setting up the National AI Portal (INDIAai) as a means to facilitate research and make the AI tools accessible to a larger number of users. The Ministry of Agricultural Affairs and ICAR, among others, have, however, devoted the concentration of AI use for climate resilience and precision farming support.

As an example, AI has enabled tech precision agriculture that is able to drastically lower input wastage and raise output by, for instance, sticking to irrigation schedules more accurately, monitoring soil conditions, and forecasting weather. Indian Agricultural Research Institute (IARI) has also implemented AI in pest detection and soil health monitoring besides other areas. CropIn and Gramophone, to name a few, are startups providing AI-driven real-time advisory services to farmers. Microsoft and

Table 2 Overview of AI governance pillars and legal frameworks in China

Governance pillar	Legal instruments	Key requirements and considerations
Content moderation	Cybersecurity law (2017)	Mandates labeling of AI-generated content (e.g., deep synthesis)
	Administrative measures for internet-based information services (2011)	Requires security assessments and filings with CAC (cyberspace Administration of China)
	Network information content ecology governance provisions (2019)	Enforces transparency in AI services prohibits illegal/ undesirable content while promoting state-approved material
	Deep synthesis provisions (2022)	Applies broadly to online platforms and AI service providers
Data protection	Personal Information Protection Law (PIPL, 2021)	Requires lawful basis and user consent for personal data processing
	Data security law (DSL, 2021)	Grants users' rights to access, correct, or delete personal data
	Recommendation algorithm provisions (2021)	Prohibits discriminatory algorithmic profiling (e.g., behavioral advertising)
	Deep synthesis provisions (2022) generative AI measures (2023)	Adheres to principles of legality, sincerity, transparency, and accountability
Algorithmic governance	Recommendation algorithm provisions (2021)	Bans algorithmic discrimination based on protected characteristics
	Deep synthesis provisions (2022)	Mandates security assessments for high-risk algorithms (e.g., generative AI)
	Generative AI measures (2023)	Requires algorithmic filings with CAC
	Ethical review measures for science and technology (trial)	Emerging focus on transparency in training data, parameters, and decision-making processes

ICRISAT worked together to create an AI-based application for sowing advisory aimed at enhancing productivity and conservation of water.

AI has been widely used in various water resource management departments. In programs like Jal Shakti Abhiyan, AI was utilized for climate prediction, irrigation scheduling, and mapping of groundwater. To make early warning and disaster preparedness through flood forecasting more effective, the Central Water Commission has adopted AI methodologies. Apart from that, AI is largely a progressive tool that can be instrumental in speeding up the transition of India to a clean energy future by 2070. AI-driven smart grid applications assist in energy distribution management while ensuring the integration of intermittent renewable resources such as solar and wind. For grid stabilization, IIT Madras developed models for forecasting solar energy generation. Implementing AI under infrastructure asset management is ReNew Power and Tata Power. The Ministry of New and Renewable Energy (MNRE) has financed such initiatives as well, where AI is utilized to balance energy loads and control storage systems.

AI has aided in another ecological activity. The monitoring of deforestation and illegal land-use activities is done by analyzing satellite images from the NRSC. The real-time monitoring of urban air quality using AI applications by the Central Pollution Control Board (CPCB) has enabled rapid response of pollution events. The adaptation of AI models by IMD has brought visible progress in terms of cyclone forecasting accuracy. AI flood early warning systems have now been collaboratively deployed in several vulnerable regions with tech companies around the globe. AI systems are proposed for use by smart cities to help increase urban resilience. For instance, any AI system that is integrated into solid waste management for the purpose of sorting and recycling, as well as the management of intra and inter-city traffic to lessen emissions and congestion in Bangalore and Delhi.

However, these changes raise several issues, including a lack of AI skills in the countryside, doubt about the infrastructural situation, biased data, difficulty algorithms to be understood, and privacy of data. The issues would be solved using digital infrastructure, raising funds for AI research and capacity building, and strengthening public-private partnerships. There is an immediate potential for the future in clear environmental monitoring and carbon credit systems that use blockchain and artificial intelligence technologies. If AI-enabled climate modeling is scaled up, there are also potential synergies for environmental forecasting and policy planning. With AI being used more for the creation of inclusive, environmentally friendly, and socially responsible growth strategies, India has the potential to be a leader in the re-alignment of its digital economy with the requirements of sustainable development.

5 Open Research Directions

Artificial intelligence offers robust capabilities that can be used to promote environmental sustainability; however, implementation is a different matter altogether as it is fraught with numerous challenges. Some of these challenges are of a technical nature while others are non-technical ones such as moral dilemmas, ambiguous rules, limitations of technology, and disparities in society. If AI is not well managed by the authorities, it could deepen the existing inequalities, use a lot of energy, or cause problems with the security of data that can make the whole situation worse, even if there are great initiatives going on in the fields of climate modeling, renewable energy, or biodiversity conservation. One of these issues is algorithmic bias, which often relies on historical data for the output and can impact vulnerable populations in a negative way when assessing climate risks or allocating resources [25]. Solutions involve equitable ML, bias reduction, and external audits. Transparency in AI decision-making is a second significant challenge. "Black box" deep learning systems are uninterpretable, an issue for carbon emission monitoring and climate prediction where explanation is necessary. Lack of justification diminishes system performance [26]. Methods such as SHAP and LIME can improve transparency for well-informed decisions. Accountability is also important. Failures, such as a wildfire forecasting system in 2021, emphasize the importance of liability frameworks establishing roles and audit trails for AI-driven decisions in sensitive domains.

The lack of international standards makes regulatory cooperation more difficult. Varying national and regional AI regulations make alignment in fields such as climate action challenging. The EU AI Act's risk-based categorization contrasts with the U.S. system, resulting in compliance challenges [27]. A global harmonized framework, perhaps through the UN, is necessary to create common norms for responsible AI uptake in environmental initiatives. One of the most difficult issues that data protection laws and privacy guidelines is the challenge of compliance and regulations associated with them. The use of AI for the sake of the environment usually involves obtaining data from personal gadgets, satellites, and IoT sensors, and therefore, there are concerns related to privacy. Privacy safeguards such as federated learning and rigorous data protection rules (e.g., GDPR) have to be put in place. Technical limitations also constrain AI in sustainability. Training large models requires a lot of energy, creating a paradox (GPT-3 emissions). Effective AI systems, compact models, pruning, distillation, and low-power hardware (neuromorphic chips) are among the possible solutions [28].

Data quality and scarcity are also major technical challenges. In many cases of sustainability, partially or incorrectly measured datasets can cause results that are not reliable. However, this problem can be solved by generating and broadening synthetic data. It is very important to motivate the creation of open-source sustainability datasets. Economic disparities between different social groups widen even more the gap of AI accessibility. Poor countries, where the lack of infrastructure and low technical capacity limit the use of AI for sustainability, are the ones to give rich countries an advantage and to leave poor communities at a disadvantage.

The elimination of this difference will need concerted global actions through the building of infrastructure and provision of capital. Figure 3 depicts the interaction of social, personal, and technological factors that lead to the creation of AI systems that are trustworthy and ethical. The swift development of AI is a typical issue for its governance. Current governance lags the development of quantum computers, generative models, and neuro-symbolic systems. AI used directly without human supervision may have unanticipated negative effects. It is necessary to incorporate technological changes into adaptive governance frameworks [29]. This involves approaches like risk-based regulation, setting up multidisciplinary ethics committees, and continuously updating legal frameworks.

To summarize, AI is a powerful tool that can significantly contribute to the realization of Sustainable Development Goals, but its efficient use requires careful consideration and appropriate regulation. If we want AI to be a source of positive impacts on a global scale including the economy, society, and environment over the next decades, it is important to strengthen institutions, modernize laws, and put in place the necessary data and technical infrastructure.

6 Conclusion

AI can be a major factor in enhancing climate models and increasing agricultural productivity, just to mention a few aspects of sustainability. It can process very complex environmental data on a large scale and in a fast manner. This is very helpful for making better decisions concerning sustainable development. However, this capability presents ethical and governance issues, such as global inequalities in access to AI and computational environmental costs. Current regulatory regimes capture both advances and significant gaps. Technical issues like data bias and the need for energy must be addressed considering urgent climate needs. The AI focus in the West risks widening global imbalances and further widening the existing disparities between developed and developing nations. To treat AI responsibly, three things are essential: global cooperation in ethical usage, energy-efficient AI systems, and access fairness for poor nations. These measures, complemented by periodic impact evaluations, may facilitate aligning sustainability goals with AI breakthroughs. Yet, if there are no flexible oversight and collaboration across different fields, AI might aggravate the issue it aims to alleviate. The choices about AI regulation being made at present will create the foundation for next sustainability projects.

References

1. N. L. Rane, S. P. Choudhary, and J. Rane, "Artificial intelligence driven approaches to strengthening Environmental, Social, and Governance (ESG) criteria in sustainable business practices: a review." Online. Available: https://ssrn.com/abstract=4843215

2. B. Mahesh, "Machine Learning Algorithms - A Review," International Journal of Science and Research (IJSR), vol. 9, no. 1, pp. 381–386, Jan. 2020, doi: 10.21275/art20203995. https://doi.org/10.21275/art20203995.
3. Y. Bengio, I. Goodfellow, and A. Courville, *Deep Learning*, 2015.
4. M. Campbell, A. J. Hoane, and F. H. Hsu, "Deep Blue," Artif Intell, vol. 134, no. 1–2, pp. 57–83, Jan. 2002, doi: 10.1016/S0004-3702(01)00129-1. https://doi.org/10.1016/S0004-3702(01)00129-1.
5. E. B.- Agenda and undefined 1992, "United Nations Conference on Environment and Development," iris.who.int, 1991, Accessed: May 10, 2025. Online. Available: https://iris.who.int/bitstream/handle/10665/171232/EB91_Inf.Doc-5_eng.pdf
6. J. A. Leggett and N. T. Carter, "Rio+20: The United Nations Conference on Sustainable Development, June 2012," Congressional Research Service, 2012, Accessed: May 10, 2025. Online. Available: https://sgp.fas.org/crs/row/R42573.pdf
7. "ISO/IEC 42001:2023 - AI management systems." Accessed: May 11, 2025. Online. Available: https://www.iso.org/standard/81230.html
8. "IBM watsonx.governance." Accessed: May 11, 2025. Online. Available: https://www.ibm.com/products/watsonx-governance
9. R. FOOD and H. AFFORDABLE, "Food security and nutrition in the World," 2022, Accessed: May 11, 2025. Online. Available: https://openknowledge.fao.org/server/api/core/bitstreams/69435b4d-b660-4aa4-8aea-dc5d3ecaac78/content
10. P. Shourie, V. Anand, R. Chauhan, S. Dangi, and S. Gupta, "Detection of Alzheimer's Disease Through Convolutional Neural Networks: An Analysis of Magnetic Resonance Imaging Images," in *2023 3rd International Conference on Smart Generation Computing, Communication and Networking (SMART GENCON 2023)*, Institute of Electrical and Electronics Engineers Inc., 2023, doi: 10.1109/SMARTGENCON60755.2023.10442192. https://doi.org/10.1109/SMARTGENCON60755.2023.10442192.
11. W. Holmes, M. Bialik, and C. Fadel, *Artificial intelligence in education: promises and implications for teaching and learning*, 2019. Accessed: May 11, 2025. Online. Available: https://discovery.ucl.ac.uk/id/eprint/10139722/
12. R. Vinuesa, H. Azizpour, I. Leite, et al., "The role of artificial intelligence in achieving the Sustainable Development Goals," *Nature Communications*, 2020, Accessed: May 11, 2025. Online. Available: https://www.nature.com/articles/s41467-019-14108-y
13. A. Khanna and P. Maheshwari, "Sustainable Industrial Energy Management: Multi-Agent Deep Reinforcement Learning with GA and Digital Twins," in *2024 7th International Conference on Signal Processing and Information Security (ICSPIS 2024)*, Institute of Electrical and Electronics Engineers Inc., 2024, doi: 10.1109/ICSPIS63676.2024.10812612. https://doi.org/10.1109/ICSPIS63676.2024.10812612.
14. R. Sharma, S. Dangi, and P. Mishra, "A Comprehensive Review on Encryption based Open Source Cyber Security Tools," in *Proceedings of IEEE International Conference on Signal Processing, Computing and Control*, Institute of Electrical and Electronics Engineers Inc., 2021, pp. 614–619, doi: 10.1109/ISPCC53510.2021.9609369. https://doi.org/10.1109/ISPCC53510.2021.9609369.
15. I. Gupta, S. Dangi, and S. Sharma, "Augmented Reality Based Human-Machine Interfaces in Healthcare Environment: Benefits, Challenges, and Future Trends," in *2022 International Conference on Wireless Communications, Signal Processing and Networking (WiSPNET 2022)*, Institute of Electrical and Electronics Engineers Inc., 2022, pp. 251–257. https://doi.org/10.1109/WiSPNET54241.2022.9767119.
16. R. Wolniak, "Artificial Intelligence in Smart Cities—Applications, Barriers, and Future Directions: A Review," mdpi.com, 2024, Accessed: May 11, 2025. Online. Available: https://www.mdpi.com/2624-6511/7/3/57
17. Kaushiv, K. S. Gill, R. Chauhan, S. Dangi, and D. Banerjee, "Internet of Things (IoT) Intrusion Detection Utilizing CNN and Fine Tuning It Using Various Optimization Parameters," in *2023 3rd International Conference on Smart Generation Computing, Communication and Networking (SMART GENCON 2023)*, Institute of Electrical and Electronics Engineers Inc., 2023, doi: 10.1109/SMARTGENCON60755.2023.10442765.

18. T. Hagendorff, "The Ethics of AI Ethics: An Evaluation of Guidelines," *Minds and Machines*, vol. 30, no. 1, pp. 99–120, Mar. 2020, doi: 10.1007/s11023-020-09517-8.
19. F. Morandín-Ahuerma, *Recommendation of the OECD Council on Artificial Intelligence: Inequality and Inclusion*. Online. Available: https://www.researchgate.net/publication/374508986
20. "Recommendation of the Council on," 2025. Online. Available: http://legalinstruments.oecd.org
21. I. Ulnicane, "Artificial intelligence in the European Union: Policy, ethics and regulation," in *The Routledge Handbook of European Integrations*, Taylor and Francis, 2022, pp. 254–269, doi: 10.4324/9780429262081-19.
22. R. Reddy et al., "From Silicon Valley to the World: U.S. AI Innovations in Global Sustainability," 2023. Online. Available: https://siliconvalley.onl
23. J. Cheng and J. Zeng, "Shaping AI's Future? China in Global AI Governance," *Journal of Contemporary China*, vol. 32, no. 143, pp. 794–810, 2023. https://doi.org/10.1080/10670564.2022.2107391.
24. V. Marda, "Artificial intelligence policy in India: A framework for engaging the limits of data-driven decision-making," *Philosophical Transactions of the Royal Society A*, vol. 376, no. 2133, Nov. 2018, doi: 10.1098/rsta.2018.0087.
25. Z. Obermeyer, B. Powers, C. Vogeli, and S. Mullainathan, "Dissecting racial bias in an algorithm used to manage the health of populations," *Science*, vol. 366, no. 6464, pp. 447–453, Oct. 2019, doi: 10.1126/SCIENCE.AAX2342.
26. C. Rudin, "Stop explaining black box machine learning models for high stakes decisions and use interpretable models instead," *Nature Machine Intelligence*, 2019, Accessed: May 11, 2025. Online. Available: https://www.nature.com/articles/s42256-019-0048-x
27. L. Floridi et al., "AI4People—An ethical framework for a good AI society: opportunities, risks, principles, and recommendations," *Minds and Machines*, vol. 28, no. 4, pp. 689–707, Dec. 2018, doi: 10.1007/S11023-018-9482-5.
28. P. Henderson, J. Hu, J. Romoff, E. Brunskill, D. Jurafsky, and J. Pineau, "Towards the systematic reporting of the energy and carbon footprints of machine learning," *Journal of Machine Learning Research*, vol. 21, pp. 1–43, 2020, Accessed: May 11, 2025. Online. Available: http://www.jmlr.org/papers/v21/20-312.html
29. A. Jobin, M. Ienca, and E. Vayena, "The global landscape of AI ethics guidelines," *Nature Machine Intelligence*, 2019, Accessed: May 11, 2025. Online. Available: https://www.nature.com/articles/s42256-019-0088-2
30. J. Zeng, "Artificial intelligence and China's authoritarian governance," *International Affairs*, vol. 96, no. 6, pp. 1441–1459, 2020, doi: 10.1093/ia/iiaa172. https://doi.org/10.1093/ia/iiaa172.

AI and Data Analytics in Biodiversity Conservation: Insights from Research and Applications

Sachi Choudhary

Abstract The loss of biodiversity poses a threat to ecosystems, which calls for technological solutions to the problem of conservation. This chapter examines the revolutionary role that artificial intelligence (AI), and data analytics play within the context of the conservation of species and habitats. This article provides a summary of previous research as well as applications in the real world. It demonstrates how these technologies contribute to the preservation of biodiversity. Leading technologies, such as machine learning, analyzes the massive information obtained from camera traps, satellites, and citizen science platforms in order to map the distribution of species, evaluate the suitability of habitats, and forecast the dangers posed by climate change and human activities. Real-time data analysis, which includes monitoring activities such as poaching and deforestation, also offers a prompt response to environmental concerns. In addition, the chapter discusses issues such as the veracity of data, ethical considerations, and the accessibility of technology in remote locations, respectively. The purpose of this chapter is to demonstrate how AI and data analytics are driving effective conservation methods by analyzing previous implementations.

Keywords Biodiversity conservation · Artificial intelligence · Data analytics · Species distribution · Real-time monitoring

1 Introduction

Biodiversity is an important aspect of healthy ecosystems, food webs, and ultimately to the health of human society. The world is losing biodiversity through habitat loss, climate change, and human activities. Artificial intelligence (AI) and data analytics will modernize the approach of biodiversity conservation and intervention platforms to better and more efficiently and more scalably monitor, analyze, and manage ecosystems. AI provides enhanced species identification, habitat monitoring, threat identification, and resource allocation to increase the effectiveness and data-driven approach

S. Choudhary (✉)
School of Computer Science, UPES, Dehradun, India

A. Khanna et al. (eds.), *Artificial Intelligence for Sustainable Development Goals*,
https://doi.org/10.1007/978-3-032-28217-0_3

Table 1 Key applications of AI and data analytics

Application area	AI/data analytics role	Citations
Species identification	Automated analysis of images/audio for species detection	[1–4]
Habitat monitoring	Real-time mapping, deforestation detection	[3–8]
Predictive modeling	Forecasting ecological changes, disaster risk	[1, 3, 4, 6, 8, 9]
Resource management	Data-driven decision-making, efficient resource use	[1, 4, 8]

of conservation action, Table 1. Data analytics and AI will change the biodiversity conservation efforts by enabling smarter, faster, and bigger conservation strategies. However, all forms of biodiversity conservation technologies pose ethical, technical, and data issues to consider in order to use them responsibly and effectively to protect climate-based biodiversity worldwide.

As an example, camera traps from forests in India, used AI to locate tigers, allowing researchers to quantify the tiger populations. These are not just limited to whatsapp group sites for short blips of action, but also allow data analytics on predicting risk to biodiversity, for instance, identifying areas in Africa that elephants may have lost their habitat because of deforestation. Drones with AI are used for monitoring coral reefs in the Great Barrier Reef area of Australia too, using existing drone capabilities to monitor for damage or potential bleaching events to coral reefs. AI, even these analytics systems, allows researchers to monitor illegal activities such as poaching on real-time data and then if at times require to analyze the patterns of the data.

1.1 Importance of Technology in Addressing Biodiversity Loss

Technology provides an enormous benefit to biodiversity conservation because it allows conservation work to happen much quicker and more effectively. "Technology" is a broad term that includes AI, drones, analytics, and other technologies that monitor animals and their environments in a reactive way and react to them in real time, meaning that more situations can be discovered before they happen (Fig. 1). For instance, AI cameras can instantly detect species, say a tiger or a pig, among animals, birds, or environmental images when the images are being taken.

Technology can improve threat prediction, i.e., where are the forests going to disappear, and allow planning for responses to those threats. Technology can also make more timely illegal action discovery. Technology can also help save scientists from having to make decisions by managing large amounts of data to come up with reasonable answers to questions and break it down into smaller datasets. Technology can help restore damaged ecosystems, i.e., farmer restoration of forests and involved

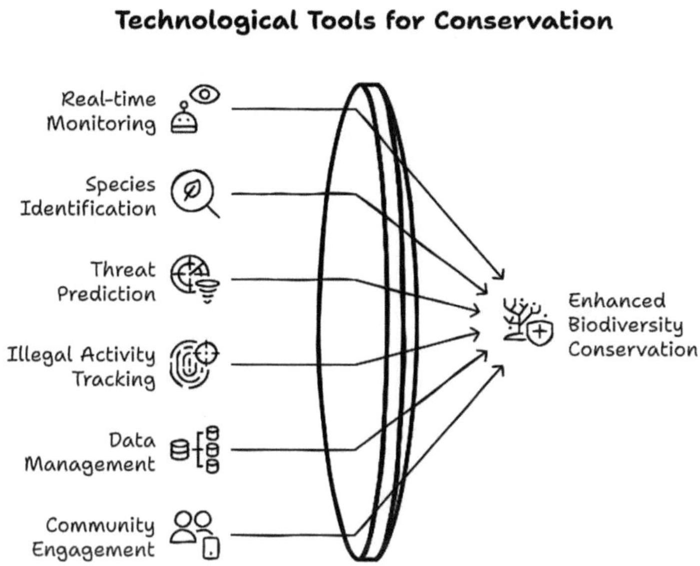

Fig. 1 Technological tools for conservation

users through applications and provide better data and save in resources and time, so that conserving nature can be easier.

In this chapter, we review the implications of data analytics and AI in relation to biodiversity conservation and how, together, they transform the field of conservation. They help with everything from monitoring ecosystems and poaching prevention to ripping down administrative burdens and creating more effective conservation. Likewise, they pose challenges, such as maintaining accuracy and ethical concerns. Real-life applications, such as how individuals are using AI to monitor jaguars in the Amazon or forecast migratory bird movements in North America, illustrate the impact of data analytics and AI. Combining technology with conservation can help us further conservation of our planet's diverse flora and fauna for generations to come.

2 Monitoring and Mapping Biodiversity

It has become increasingly possible to monitor and document biodiversity through a range of technologies, including remote sensing, satellites, AI-based identification, drones, and Internet-of-Things (IoT) sensors, Table 2. They allow for fast and precise evaluation of species and habitats at the scale for which they are subject to conservation and management. Figure 2 showcases the process to use technological tools for biodiversity monitoring and mapping.

Table 2 Technology and methods for monitoring and mapping biodiversity

Technology/method	Application area	Key outcomes/advantages	Citations
Remote sensing/satellites	Habitat mapping, ecosystem monitoring	Large-scale, high-resolution, multi-biome coverage	[10–15]
AI (image/sound)	Species identification, change detection	Automated, accurate, and rapid biodiversity assessment	[7, 16, 17]
Drones	Real-time habitat/species monitoring	Flexible, high-resolution, access to remote/sensitive areas	[7, 10, 13, 14, 17]
IoT devices	Continuous environmental/species data	Real-time monitoring, early threat detection, supports rapid response	[10, 14]

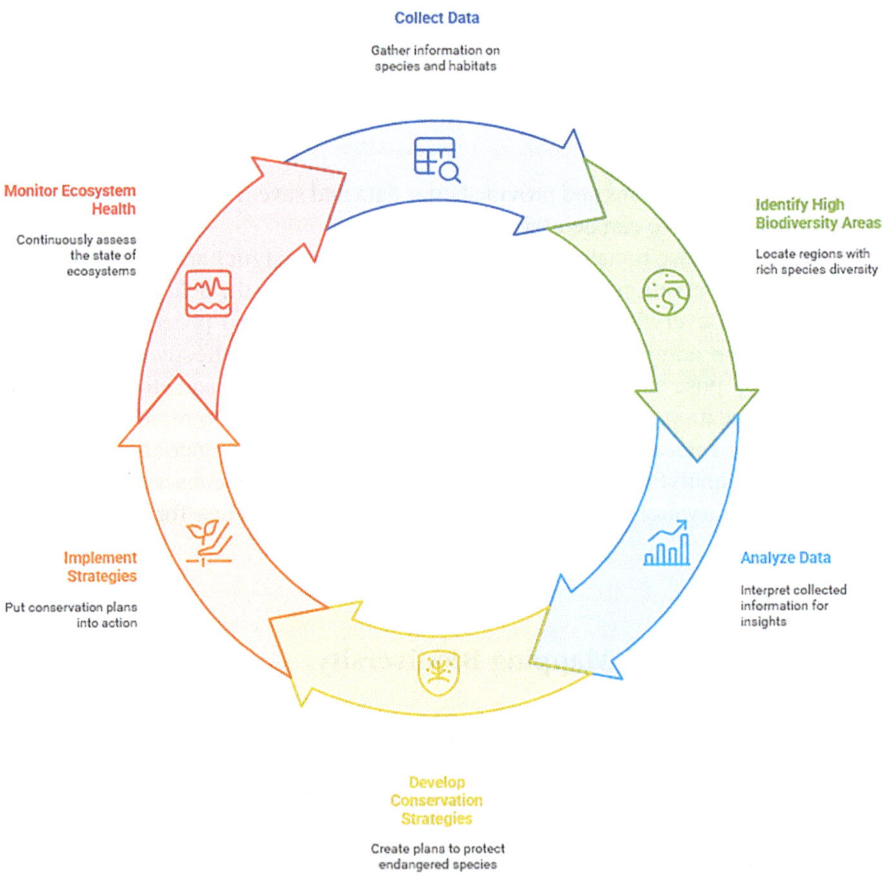

Fig. 2 Biodiversity monitoring and mapping cycle

2.1 Remote Sensing and Satellite Imagery for Habitat Mapping

Remote sensing and satellite imagery are used by scientists to examine large land and ocean areas in detail. The technologies can be employed to document an area's habitat, follow forests over time, determine instances of deforestation, and now follow the trends in an area's ecosystem over time. They work in both natural and managed situations and are used for knowing the health of ecosystems [10–15]. Scientists also employ satellite data to document indicators of key biodiversity features, many of which are related to/represented by vegetative features such as plant structure, plant species, and habitat connectivity or fragmentation. The indicators are labeled as Essential Biodiversity Variables (EBVs) and are important to the study and conservation of the natural world. With improved technology, and improved satellite sensors, scientists are able to acquire more information, in more detail, and more regularly [12, 13]. By combining different types of images and instruments, scientists can achieve more accurate descriptions of biodiversity and how it is changing to either facilitate or support conservation [12–14].

2.2 AI-Driven Species Identification

Currently, artificial intelligence (AI) is widely used to analyze images and sounds in nature. It helps analyze images from satellites and drones, and sounds from forests or oceans, and allows identification of animals and follow where they go, all quicker and easier than manually doing it [7, 16, 17].

This is done with deep learning, a type of AI, that recognizes small changes in nature, and tracks how species move or evolve. AI interfaces with the remote-sensing instruments and helps create maps of where species are located, particularly in areas that are hard to get to. It also looks at changes in biodiversity using drone data [16, 17]. There can be a few problems with this. The quality of information is not always there, and the AI platforms typically need a lot of samples to learn. Also, in some areas, trained people were needed to work with and understand how to use these tools—that is a challenge [16, 17].

2.3 Drones and IoT Devices for Real-Time Biodiversity Monitoring

Drones are an incredible technology that can view the environment from above. Drones can travel over forests, mountains, or those places where you can't normally take a camera to capture high-resolution video or images of a specific location [7, 10, 13, 14, 17].

The drones allow us to look at things in the environment as monitoring habitats, tracking species, and even seeing illegal activity such as logging or poaching. We can also use IoT devices, like smart sensors, on the ground that collect ongoing environmental (i.e., temperature, moisture), and presence data (i.e., species) of some animal species with sensors [10, 14]. The devices give us the opportunity to understand and respond to situations that need action to protect the environment. IoT devices, drones, and AI technologies work together to create an intelligent network capable of monitoring conditions and supporting informed decision-making. In short, with developments in drone technologies, IoT devices and AI, we have a much more efficient way to measure and observe the biodiversity of the planet [10, 14, 17].

3 Predictive Modeling for Conservation Planning

Especially through machine learning (ML), predictive modeling is becoming a new basis for conservation planning, supporting a much more accurate estimate of species distributions, habitat suitability, and the impacts of climate and human activity. Random forest is one example of an ML algorithm that consistently outperforms traditional species distribution modeling (e.g., SDMs) with both accuracy and reliability, as demonstrated by research suggesting that the sage-grouse's current and future habitats can be predicted, Table 3. These models provide a comprehensive suite of data about climate predictors, vegetation cover, and species locations in that region to provide future potential facilitating habitat and ability to project alternative scenarios due to projected temperature change. For example, a study modeling sage-grouse habitat in Utah with an ML-based prediction would suggest a highly fragmented potential future habitat, with upwards of 70% of available habitat lost due to a climate change impact. This emphasizes a need for a focus on connectivity and adaptation-focused conservation approaches [18].

In addition to species distribution, machine learning (ML) models now provide quantitative assessments of the impact of human activity and land-use planning on natural environments. When informed by interdisciplinary datasets, ML can predict

Table 3 Applications of predictive modeling for conservation planning

Application area	Predictive modeling role	Citations
Species distribution	Identifies current/future habitats, guides connectivity	[18–20]
Habitat suitability	Assesses area suitability for conservation	[19, 21]
Climate change impact	Forecasts habitat/species shifts, informs adaptation	[18, 19, 22–25]
Human activity impact	Simulates land-use effects, supports mitigation	[21, 22, 26]
Risk assessment	Quantifies vulnerability, supports resource allocation	[18, 20, 23, 24]

the consequences of planning decisions (e.g., soil sealing) and contribute to nature-based solutions at local/regional scales. The development of such predictive capacities informs spatial decision-makers as they think through the longer-term implications for development and potentially move toward more sustainable land-use policies.

In addition, ML models are also being utilized in predicting a series of ecological changes (e.g., including, but not limited to deforestation, air pollution, sea level rise, and temperature change) by identifying relationships in complex ecological systems. These predictive capabilities support land-use planners and conservation practitioners in making informed decisions, developing effective conservation strategies, and allocating resources more efficiently for biodiversity protection [22, 26].

Risk is also a relevant use of predictive modeling within the sphere of conservation. Predictive models based in ML can quantify the exposure of susceptible species and populations to habitat loss, population decline, and/or extreme event-based occurrences. As an example of intended spatial extrapolation risk assessment, the models will enable a system to evaluate the soundness of the predictions (e.g., based in ML) and/or to create the reliable predictions that might serve as a trigger for their next action [24].

Likewise, iterative spatial portfolio analysis facilitates the deployment of resources for conservation even in the absence of climate change data, thereby minimizing the expected loss in conservation outcome. These initiatives provide forward momentum; nevertheless, there are problems yet to overcome including accessibility to high-quality, standardized data; interpretable, usable models; and meaningful modularization of prediction in formal policy frameworks. Interdisciplinary collaboration and expert knowledge are needed to optimize the use of predictive models in planning for conservation [19, 22].

3.1 Data Integration and Management

Data integration and management are necessary in support of biodiversity data in light of and the large increase in the quantity and variability of biodiversity data over the past few years. Integrating and assimilating information from various sources, including community-based science projects, collections of museum specimens, remote sensing, and digital online, provides a wider lens with which to view biodiversity at taxonomic, spatial, and temporal resolution. Projects such as Global Biodiversity Information Facility (GBIF), for example, opened new avenues for open-access aggregation of data and integration of biodiversity information from all over the globe. Aggregating data across disciplines of environmental science, evolutionary biology, conservation, and policy, facilitates studies using varied aspects of biodiversity. The availability of an integrated and mobilized biodiversity data system, and data in general, has risen, enabling both basic and applied research and providing opportunities for analyses of the data that are not available in unintegrated and fragmented data [27].

Nonetheless, integration and mobilization are impeded by heterogeneity across the type of data, spatial or temporal scale of observation, purpose of data collection, and terminology. Addressing this heterogeneity can be achieved by first identifying the data domain (i.e., what aspect of biodiversity is described) and then the informational resolution (i.e., the precision of the bass information). There are a variety of approaches to the integration and mobilization of data including developing standardized methods of data collection and mobilization, using imputation methods to fill gaps, and developing open-access common repositories for data sharing [28].

Considering both the domain and resolution of the data enables policies to identify biases and gaps so macroecological studies are more representative and inclusive. Integrated biodiversity data also are evaluated systematically to look for gaps in management and conservation (for instance—species with no representation, or regions without any management actions). Integrated frameworks look for mismatches in the type of resources available as well as information about pressures on the environment and applied conservation actions to help set priorities for conservation action [29]. Moreover, although challenges remain in providing adequate metadata coverage and assessment quality, the use of automation and sophisticated algorithms provides opportunities for streamlined accessing, evaluating and integrating large biodiversity information management systems. Cultivating a culture of integration through collaboration is essential: scientists, policy actors, communities, and data managers, in that order. The networked system fosters resilience and builds joint ownership of monitoring and management of biodiversity. The connectivity of the system and actions to integrate biodiversity information is promoted by actors' scientific and societal reason for engaging in formal biodiversity reporting and management [30]. Appropriate best practices for integrated biodiversity information data management emphasize the importance of adequate metadata, long-term persistence, and reproducibility in open access and peer reviewed platforms while creating opportunities for further analysis and use. Ultimately, it is essential that sustainable practices of aggregating and managing biodiversity information are able to inform decision making leading to the best possible conservation outcomes.

3.2 AI in Wildlife Protection and Anti-poaching Efforts

Artificial intelligence (AI) is leading a revolution in wildlife protection and poaching prevention through real-time monitoring, prediction analysis, and automated threat detection, Table 4. AI platforms, such as Conservation AI, combine machine learning with computer vision to detect and classify animals, people, and poaching-related objects. This function can be performed via visual or thermal cameras for immediate detection or classification of potential threats. AI-based models incorporate huge amounts of data using neural networks, leading to immediate alerts during time-constrained situations such as poaching, while also aiding in long-term monitoring of species and habitats. The integration of AI capabilities with drones and IoT technologies creates an even greater level of surveillance, offering the ability to monitor large areas with reduced human interaction allowing for the easy recognition of illegal activity in even the largest or most remote of spaces. For example, AI-based monitoring systems have been used for tracking poaching activity, to predict poachers' movements, and can alert enforcement to suspicious events, all while using predefined logic or algorithms. This offers wildlife conservation and anti-poaching projects with a proactive approach that surpasses conventional capabilities [31, 33, 34].

Machine learning pipelines, like the Protection Assistant for Wildlife Security (PAWS) [32], have been developed to quantify poaching risk by identifying high-risk poaching areas and optimize patrol routes for law enforcement. A number of obstacles

Table 4 Applications of AI in wildlife protection and anti-poaching

AI application/ platform	Key features and technologies	Impact on wildlife protection and anti-poaching	Source
Trail-tracker	Real-time monitoring using AI and IoT; tracks poaching activity; predicts poacher behavior; alerts authorities	Enables rapid response to poaching threats; prevents illegal activities	[31]
Protection assistant for wildlife security (PAWS)	Machine learning pipeline; identifies high-risk poaching areas; optimizes patrol routes; uses Gaussian processes for uncertainty	Increases detection of snares by 30%; robust patrol planning; field-tested in multiple parks	[32]
Conservation AI platform	Machine learning and computer vision; visual and thermal cameras; CNNs and transformers; real-time and long-term monitoring	Detects/classifies animals, humans, and poaching objects; supports global biodiversity monitoring	[33]
Smart conservation	AI for automated species identification (image/audio); habitat health assessment; predictive modeling; real-time alerts	Improves monitoring, resource allocation, and community engagement; proactive anti-poaching	[4]
Drones + AI	Drones with cameras, LiDAR, thermal imaging; AI for data analysis, anomaly detection, species ID, habitat mapping	Monitors large/inaccessible areas; real-time detection of poaching and habitat threats	[34]
AI-driven behavior monitoring	Computer vision, deep learning (YOLOv8); real-time species/ behavior detection; anomaly alerts	Tracks animal behavior, migration, and distress; triggers alerts for poaching	[35]
AI for environmental monitoring	AI analyzes camera trap, drone, GPS, and acoustic data; species ID; population estimation; habitat assessment	Enhances anti-poaching, biodiversity analysis, and early threat detection	[3]
Wildlife watch	Deep learning for individual animal ID; integrates camera, drone, bioacoustics, and environmental sensors	Enables real-time tracking, early poaching detection, and habitat threat prediction	[36]

have to be overcome, such as data imbalance and uncertain issues, for these systems to develop patrol routes. Some of the techniques used include Gaussian processes to increase the robustness of patrol planning and improve the detection of poaching tools such as snares. In field studies conducted in national parks, PAWS demonstrated an ability to significantly improve anti-poaching operations, with higher detection rates and adaptability in numerous environments.

AI was found to add value in the automation of species identification, recording ecological behavior, and aggregating data on animals from varied sources, all contributing to a diligent managing process based on evidence and resource optimization. In particular, environmental predictive modeling can be enabled through historical data and real-time observations which can lead to forecasts of what is expected, guiding the conservation practitioner toward preventive actions in line with their conservation goals, while creating the potential for better protocols to be developed. However, while the potential of applied AI for conservation is vast, efficacy will be achieved only through an ethical deployment of technology in conservation. This must consider context-specific issues on data privacy, and algorithm-based visibility, and equitable access, ensuring communities, including people and wildlife, both benefit from technological advancements.

In summary, AI provides solutions to wildlife protection and anti-poaching in scalable, adaptable, and efficient ways. The use of AI for wildlife protection has enhanced the global development and progress against illegal wildlife crime, reinforcing our growing global commitment to preserving biodiversity [4, 34, 35].

4 Challenges and Ethical Considerations

The intersection of AI and data analytics for biodiversity protection offers exciting and dynamic opportunities along with challenging implications and ethics, Fig. 3. Data privacy and the ethical use of AI in conservation are critical considerations, as AI-driven conservation initiatives present substantial potential benefits while also introducing risks, complexities, and concerns related to governance, transparency, and responsible data management [9]. As AI systems begin to rely on ever-increasing amounts of data collected from sensors, remote monitoring, and citizen science, protecting or preserving people's, groups', ecologies, and/or vulnerable places' privacy concerns is paramount. There is an articulated need for policy frameworks that go beyond the notion of "data sovereignty" to account for the political and historical legacies surrounding issues of data ownership [37, 38]. Policy frameworks should consider issues of data privacy and security of sensitive ecological data and sensitive metadata, there needs to be clear policies to ensure the transparency of algorithms and equitable access to the use of different types of AI, wants and desires of communities and regions that are benefitting from these types of technologies. Ethical guidelines regarding the collection, storage, and use of biodiversity data should also be considered, especially in regard to illegally held Indigenous and other peoples' land or space or sensitive habitat issues. Further, a serious problem with AI model algorithm's capabilities for biodiversity monitoring, documentation, or decision making is that they may further operate to amplify, duplicate and replicate existing biases that may be present in the data used to train AI models. So, problematic AI models will perpetuate biases through AI models to realize conservation outcomes but can also lead to failure in recognition of species, findings that are not representative due to the over-extended biases in the model, or bad decisions regarding protection for species and ecosystems. These considerations consider data quality, data standardization, establishing validation of models. These current considerations are their value not only as reflections of uniqueness but also as our best ways of working toward elimination of social biases recorded and perpetuated in AI systems from the earliest days of AI coding and development [37, 38].

Interdisciplinary collaboration and stakeholder involvement are important for designing ethical AI solutions that are scientifically sound and socially accountable [9]. It is also important to create equilibrium between AI and technological intervention and human-based conservation. AI and data analytics can improve species identification, habitat use monitoring, and natural resource prioritization when mapping and planning conservation work; however, too much reliance on a technological process can lead to excluding traditional ecological knowledge and community-based conservation practices, making the long-term goal of biodiversity and conservation preservation less achievable. Preserving biodiversity is challenging and work should be based on a collaborative synthesis of modern technology and traditional human practices and value systems. If practitioners continue to prioritize and better incorporate modern technology in differing forms in conservation practices, local knowledge, histories, practices, processes, stories, and values need to be integrated socially responsible and symbolically accurate into that decision making process [1, 7]. Community engagement through place-based participatory governance frameworks promotes inclusive decision-making and facilitates the integration of local knowledge and stakeholder perspectives into conservation planning and implementation.

In summary, the responsible use of AI and data analytics in the name of biodiversity preservation lies with some ethical frameworks in mind, attempting to mitigate

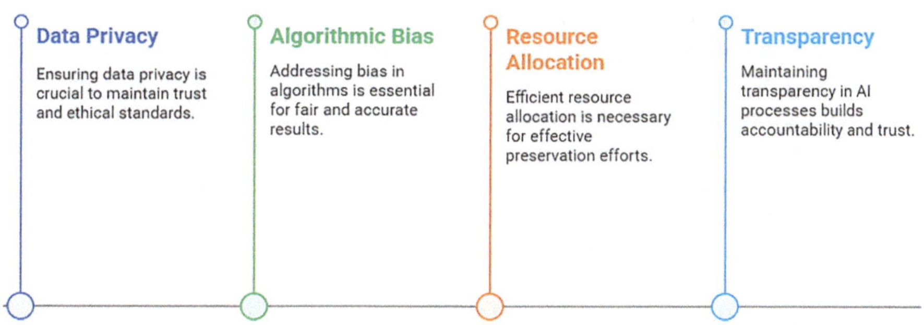

Fig. 3 Challenges and ethical consideration on implementation of technologies in biodiversity

any algorithmic bias and that the use of technologies may seem innovative or an efficient decision-making process but we are primarily referring to our experiences with indigenous, first nation and traditional ecological knowledge. Finding innovative pathways that provide equity for all preserving returning to modal practices may be challenging. Addressing the issues and challenges outlined above is the underlying basis for generating effective, relevant, and equitable conservation solutions in BC.

Case Studies and Real-World Applications

The application of AI-enabled conservation projects has had great success in a variety of areas, including water management, biodiversity conservation, and forest protection. For example, AI-enabled solutions such as ChatGPT allowed for real-time conservation issues hand feel better for water conservation because of better, more predictive insights, better resource efficiency, and precision irrigation. There are other projects such as Klir Comply and Rammas, which allowed water managers and compliance professions to improve water quality control and the underlying decision-making involved in resource management through AI-enabled solutions [39]. Likewise, AI has emerged as an important technology in biodiversity conservation by enhancing habitat monitoring, improving species identification, assessing threats to ecosystems, and generating accurate data that supports effective conservation planning and improved conservation outcomes. The case studies used by the integrated conservation initiative illustrate how machine learning algorithms and other AI-based and data-based technologies have emerged to help address current conservation issues and how to integrate traditional conservation projects with modern technologies to achieve more coherent and adaptable conservation solutions [1].

Global initiatives have imparted useful lessons, highlighting the need for ethical oversight, transparency, and stakeholder engagement. Successful initiatives illustrate the importance of governance in responsible AI use, particularly in terms of data privacy and equal access to AI technologies [39, 40].

The incorporation of big data and AI in the United States, for example, has improved decision-making and management practices, demonstrating how these technologies can transform the relationship between economic and ecological activity to the protection of biodiversity. Additionally, AI strategies to reduce deforestation have achieved an impressive rate of accuracy concerning identifying sites of deforestation compared to more conventional approaches [10], allowing for a more targeted approach to mitigating deforestation and land use, prioritizing substantial areas of concern such as the Amazon Basin, Central Africa, and supply chains (wood and palm oil) based in Southeast Asia.

AI solutions can be consistently scaled or modified to fit new campaigns and projects toward conservation; this flexibility in AI technologies supports the connection of solutions to various ecosystems and ecological issues, as well as the ability of ecosystems to be adopted in other geographical areas or ecosystems [3].

Tools for decision-making powered by AI can provide enormous possibilities in conservation planning and resource allocation, while enhancing prediction accuracy and identifying specific interventions. For instance, when managing urban biodiversity, AI's accuracy with tracking wildlife has outperformed 90% of predictions, while improving what was considered an accurate prediction to become 18.5% accuracy better than what existed by classical methods [38]. These uses will lead to improved prioritization and enhanced effectiveness to any conservation with limited resources. AI can also be combined with classical methods to produce the best outcomes for conservation. As AI can resolve the limitations of classical approaches (e.g., there is a higher limit for scalability) and gaps in knowledge or methods that exist within traditional approaches, knowledge derived from these traditional methods can provide context-specific solutions to the problem by delivering culturally appropriate decisions. Together, these emergent approaches can allow for resilient and adaptive approaches in conservation.

Although the future of AI technologies presents promising opportunities for traditional and local conservation, various challenges to implementation exist. These challenges may include issues related to data quality and accessibility, algorithmic bias, ethical issues, and other environmentally and socially just effects that prohibit the use of AI from marginalizing local communities. Addressing these issues will highlight the need for capacity building and equitable access to AI technologies in whatever form it takes, particularly in regions of the globe that are already resource-constrained.

5 Conclusion

It is undeniable that AI and data analytics have been essential in the preservation of biodiversity. Identification of species, monitoring of ecosystems, and evaluation of threats have all become simpler and more precise due to the integration of these leading technologies. These technologies made it possible to identify species. Numerous case studies and detailed reviews demonstrate that AI-driven methodologies have achieved improvements in data management, predictive modeling, and resource allocation which has resulted in conservation solutions that are more effective and robust. Additionally, it has improved the planning for conservation and the analysis of ecosystems.

However, dealing with the implications of AI is not always an easy task. In order for it to be successful, it is necessary to address ethical concerns, ensure that data is of high quality, and ensure that everyone has equal access to technology. A significant amount of literature emphasizes the significance of integrating AI with conventional conservation techniques to achieve the most favorable outcomes and guarantee that conservation efforts will be successful throughout time. Employing artificial intelligence and data analytics is, in general, a significant step forward in the protection of biodiversity. On the other hand, in order to achieve success over the long term, we must continue to consider concerns that are both ethical and social in nature.

References

1. F. Ullah, S. Saqib, and Y. C. Xiong, "Integrating artificial intelligence in biodiversity conservation: bridging classical and modern approaches," Biodivers Conserv, Jan. 2024, doi: https://doi.org/10.1007/S10531-024-02977-9.
2. C. Li, "Biodiversity assessment based on artificial intelligence and neural network algorithms," Microprocess Microsyst, vol. 79, Nov. 2020, doi: https://doi.org/10.1016/J.MICPRO.2020.103321.
3. Onyebuchi Nneamaka Chisom, Preye Winston Biu, Aniekan Akpan Umoh, Bartholomew Obehioye Obaedo, Abimbola Oluwatoyin Adegbite, and Ayodeji Abatan, "Reviewing the role of AI in environmental monitoring and conservation: A data-driven revolution for our planet," World Journal of Advanced Research and Reviews, vol. 21, no. 1, pp. 161–171, Jan. 2024, doi: https://doi.org/10.30574/WJARR.2024.21.1.2720.
4. N. R. Hegde and Dr. S. M. Bargavi, "Smart Conservation: Integrating AI for Enhanced Wildlife Monitoring," International Research Journal of Computer Science, vol. 11, no. 01, pp. 62–67, Jan. 2024, doi: https://doi.org/10.26562/IRJCS.2024.V1101.10.
5. K. N. Shivaprakash et al., "Potential for Artificial Intelligence (AI) and Machine Learning (ML) Applications in Biodiversity Conservation, Managing Forests, and Related Services in India," Sustainability, vol. 14, no. 12, Jun. 2022, doi: https://doi.org/10.3390/SU14127154.
6. K. S. Pokkuluri, T. Mounika, N. D. Devi, D. R. Kishore, B. Balakiruthiga, and B. M. Krishna, "Disaster Management Based on Biodiversity Conservation Using Remote Sensing Data Analysis Using Machine Learning Model," Remote Sens Earth Syst Sci, 2025, doi: https://doi.org/10.1007/S41976-024-00188-Y.
7. A. K. Rathoure and B. Laxmibai GP Ram, "Unveiling the marvels of biodiversity: recent advancements in conservation efforts," Biodiversity International Journal, vol. 7, no. 2, pp. 51–61, Jul. 2024, doi: https://doi.org/10.15406/BIJ.2024.07.00211.
8. Victoria Bukky Ayoola, Idoko Peter Idoko, Samson Ohikhuare Eromonsei, Olusegun Afolabi, Akinkunmi Rasheed Apampa, and Oluwatosin Seyi Oyebanji, "The role of big data and AI in enhancing biodiversity conservation and resource management in the USA," World Journal of Advanced Research and Reviews, vol. 23, no. 2, pp. 1851–1873, Aug. 2024, doi: https://doi.org/10.30574/WJARR.2024.23.2.2350.
9. Uwaga Monica Adanma and Emmanuel Olurotimi Ogunbiyi, "Artificial intelligence in environmental conservation: evaluating cyber risks and opportunities for sustainable practices," Computer Science & IT Research Journal, vol. 5, no. 5, pp. 1178–1209, May 2024, doi: https://doi.org/10.51594/CSITRJ.V5I5.1156.
10. B. Haq et al., "Tech-Driven Forest Conservation: Combating Deforestation With Internet of Things, Artificial Intelligence, and Remote Sensing," IEEE Internet Things J, vol. 11, no. 14, pp. 24551–24568, 2024, doi: https://doi.org/10.1109/JIOT.2024.3378671.
11. M. Finizio et al., "Remote Sensing for Urban Biodiversity: A Review and Meta-Analysis," Remote Sensing, vol. 16, no. 23, p. 4483, Nov. 2024, doi: https://doi.org/10.3390/RS16234483.
12. C. S. Reddy et al., "Remote sensing enabled essential biodiversity variables for biodiversity assessment and monitoring: technological advancement and potentials," Biodivers Conserv, vol. 30, no. 1, pp. 1–14, Jan. 2021, doi: https://doi.org/10.1007/S10531-020-02073-8.
13. P. Vihervaara et al., "How Essential Biodiversity Variables and remote sensing can help national biodiversity monitoring," Glob Ecol Conserv, vol. 10, pp. 43–59, Apr. 2017, doi: https://doi.org/10.1016/J.GECCO.2017.01.007.
14. G. Ali, M. M. Mijwil, I. Adamopoulos, and J. Ayad, "Leveraging the Internet of Things, Remote Sensing, and Artificial Intelligence for Sustainable Forest Management," Babylonian Journal of Internet of Things, vol. 2025, pp. 1–65, Jan. 2025, doi: https://doi.org/10.58496/BJIOT/2025/001.
15. N. Pettorelli et al., "Satellite remote sensing of ecosystem functions: opportunities, challenges and way forward," Remote Sens Ecol Conserv, vol. 4, no. 2, pp. 71–93, Jun. 2018, doi: https://doi.org/10.1002/RSE2.59.

16. N. Pettorelli, J. Williams, H. Schulte to Bühne, and M. Crowson, "Deep learning and satellite remote sensing for biodiversity monitoring and conservation," Remote Sens Ecol Conserv, Apr. 2024, doi: https://doi.org/10.1002/RSE2.415.
17. C. A. Cabrera-Verdesoto, K. G. Arellano-Reinoso, and D. A. Ureta-Leones, "Uso de inteligencia artificial para la identificación de especies forestales a través de imágenes satelitales," Multidisciplinary Collaborative Journal, vol. 2, no. 1, pp. 26–38, Feb. 2024, doi: https://doi.org/10.70881/MCJ/V2/N1/29.
18. A. C. M. Fernandes, R. Q. Gonzalez, M. A. Lenihan-Clarke, E. F. L. Trotter, and J. J. Arsanjani, "Machine Learning for Conservation Planning in a Changing Climate," Sustainability, vol. 12, no. 18, Sep. 2020, doi: https://doi.org/10.3390/SU12187657.
19. K. A. Wood, R. A. Stillman, and G. M. Hilton, "Conservation in a changing world needs predictive models," Anim Conserv, vol. 21, no. 2, pp. 87–88, Apr. 2018, doi: https://doi.org/10.1111/ACV.12371.
20. K. J. Gutzwiller and K. M. Serno, "Using the risk of spatial extrapolation by machine-learning models to assess the reliability of model predictions for conservation," Landsc Ecol, vol. 38, no. 6, pp. 1–10, Jun. 2023, doi: https://doi.org/10.1007/S10980-023-01651-9.
21. J. Forster, S. Bindreiter, B. Uhlhorn, V. Radinger-peer, and A. Jiricka-pürrer, "A Machine Learning Approach to Adapt Local Land Use Planning to Climate Change," Urban Plan, vol. 10, 2025, doi: https://doi.org/10.17645/UP.8562.
22. E. A. Etukudoh, A. Hamdan, K. I. Ibekwe, A. A. Umoh, and V. I. Ilojianya, "AI and machine learning in climate change research: A review of predictive models and environmental impact," World Journal of Advanced Research and Reviews, vol. 21, no. 1, pp. 1999–2008, Jan. 2024, doi: https://doi.org/10.30574/WJARR.2024.21.1.0257.
23. G. Rilov et al., "Adaptive marine conservation planning in the face of climate change: What can we learn from physiological, ecological and genetic studies?," Glob Ecol Conserv, vol. 17, Jan. 2019, doi: https://doi.org/10.1016/J.GECCO.2019.E00566.
24. P. Shah, M. L. Mallory, A. W. Ando, and G. R. Guntenspergen, "Fine-resolution conservation planning with limited climate-change information," Conservation Biology, vol. 31, no. 2, pp. 278–289, Apr. 2017, doi: https://doi.org/10.1111/COBI.12793.
25. A. Batool, Z. Ali, M. Mohsin, A. Masmoudi, V. Kartal, and S. Satti, "Assessing the generalization of forecasting ability of machine learning and probabilistic models for complex climate characteristics," Stochastic Environmental Research and Risk Assessment, vol. 38, no. 8, pp. 2927–2947, Aug. 2024, doi: https://doi.org/10.1007/S00477-024-02721-3.
26. L. Kumar, A. Archana, and K. Nath, "Predicting and Navigating Environmental Transformations using Machine Learning," Procedia Comput Sci, vol. 235, pp. 889–896, 2024, doi: https://doi.org/10.1016/J.PROCS.2024.04.084.
27. J. M. Heberling, J. T. Miller, D. Noesgaard, S. B. Weingart, and D. Schigel, "Data integration enables global biodiversity synthesis," Proc Natl Acad Sci U S A, vol. 118, no. 6, Feb. 2021, doi: https://doi.org/10.1073/PNAS.2018093118.
28. C. König, P. Weigelt, J. Schrader, A. Taylor, J. Kattge, and H. Kreft, "Biodiversity data integration—the significance of data resolution and domain," PLoS Biol, vol. 17, no. 3, Mar. 2019, doi: https://doi.org/10.1371/JOURNAL.PBIO.3000183.
29. M. Lanzas et al., "Detecting management gaps for biodiversity conservation: An integrated assessment," J Environ Manage, vol. 354, Mar. 2024, doi: https://doi.org/10.1016/J.JENVMAN.2024.120247.
30. H. S. Kühl et al., "Effective Biodiversity Monitoring Needs a Culture of Integration," One Earth, vol. 3, no. 4, pp. 462–474, Oct. 2020, doi: https://doi.org/10.1016/J.ONEEAR.2020.09.010.
31. V. Zope, S. Relekar, P. Choudhary, M. Ochaney, and R. Bhagtani, "Trail-tracker: anti-poaching intelligence using AI and IoT," Proceedings of the IEEE International Conference on Image Information Processing, pp. 312–315, 2023, doi: https://doi.org/10.1109/ICIIP61524.2023.10537669.
32. L. Xu et al., "Stay Ahead of Poachers: Illegal Wildlife Poaching Prediction and Patrol Planning Under Uncertainty with Field Test Evaluations (Short Version)," 2020 IEEE 36th International

Conference on Data Engineering (ICDE), vol. 2020-April, pp. 1898–1901, Apr. 2020, doi: https://doi.org/10.1109/ICDE48307.2020.00198.
33. P. Fergus, C. Chalmers, S. N. Longmore, and S. A. Wich, "Harnessing Artificial Intelligence for Wildlife Conservation," ArXiv, vol. abs/2409.10523, doi: https://doi.org/10.48550/ARXIV.2409.10523.
34. R. Bekmurzaeva, R. Kalimullin, and F. Aguzarova, "Application of drones and artificial intelligence to monitor and protect natural ecosystems," BIO Web Conf, vol. 140, Nov. 2024, doi: https://doi.org/10.1051/BIOCONF/202414001008.
35. Pandiselvi, R., Jeyaprabhu, J., Jerome Immanuel Jebaraj, J., Muthupandi, L., "AI-driven Wildlife Behavior Monitoring using Computer Vision," International Journal for Multidisciplinary Research, vol. 6, no. 5, Oct. 2024, doi: https://doi.org/10.36948/IJFMR.2024.V06I05.29257.
36. C. Chimwemwe, "Wildlife watch," i-manager's Journal on Information Technology, vol. 13, no. 2, p. 24, 2024, doi: https://doi.org/10.26634/JIT.13.2.20819.
37. A. K. Rathoure and A. K. Rathaur, "AI and IoT in biodiversity assessment," Journal of Aquaculture & Marine Biology, vol. 13, no. 3, pp. 108–115, 2024, doi: https://doi.org/10.15406/JAMB.2024.13.00404.
38. Y. Rahmati, "Artificial Intelligence for Sustainable Urban Biodiversity: A Framework for Monitoring and Conservation," ArXiv, vol. abs/2501.14766, doi: https://doi.org/10.48550/ARXIV.2501.14766.
39. A. U. Egbemhenghe et al., "Revolutionizing Water Treatment, Conservation, and Management: Harnessing the Power of AI-Driven ChatGPT Solutions," Environmental Challenges, vol. 13, Dec. 2023, doi: https://doi.org/10.1016/J.ENVC.2023.100782.
40. J. Cavender-Bares et al., "Integrating remote sensing with ecology and evolution to advance biodiversity conservation," Nat Ecol Evol, vol. 6, no. 5, pp. 506–519, May 2022, doi: https://doi.org/10.1038/S41559-022-01702-5.

Artificial Intelligence: A Tool for Sustainable Development and Its Emerging Research Directions

Chetankumar Chudasama and Deepak Kumar Verma

Abstract Artificial intelligence (AI) has developed into a cutting-edge technology that provides solutions to a myriad of complex, interconnected socio-economic and environmental sustainability challenges. AI applications nevertheless impact and contribute to numerous UN Sustainable Development Goals (SDGs) on climate change, renewable energy interconnections, predictive and precision agriculture, integrated and equitable healthcare, innovative and sustainable cities and transportation, environmental monitoring and assessment, and systems of the circular economy. This chapter provides a structured evaluation of the role of AI in advancing sustainability, with a literature synthesis covering the years 2020–2025. This assessment covers the risks, the potential, and the purposive applications/impacts of AI. We assess domain-specific applications, energy overconsumption and bias, transparency and inequitable control/access issues in governance-challenged methodologies, and the primary enabling technologies, including but not limited to: Green AI, privacy-preserving federated and explainable AI, digital twins, and AI. The chapter concludes with suggested and emerging directions for research and policy to foster the responsible and equitable use of AI that safeguards environmental sustainability and places radical transparency on the governance of AI in relation to sustainable development.

Keywords Artificial intelligence · Sustainable development · Green AI · Federated learning · Digital twins · Explainable AI · Climate action · Smart cities

1 Introduction

In order to attain sustainable development, there needs to be integrated, balanced, and coordinated action across the social, financial, economic, and environmental domains to address the challenges that the world is currently facing. An interdisciplinary and systemic approach is needed to address the challenges facing the world today, and the 2030 Agenda for Sustainable Development is highlighting this need through

C. Chudasama · D. K. Verma (✉)
Department of Computer Engineering, Marwadi University, Rajkot, Gujarat, India

A. Khanna et al. (eds.), *Artificial Intelligence for Sustainable Development Goals*,
https://doi.org/10.1007/978-3-032-28217-0_4

its 17 Sustainable Development Goals (SDGs). In this scenario, machine learning (ML), deep learning (DL), reinforcement learning (RL) and artificial intelligence (AI), as advanced data processing technologies, are promising highly influential, and transformative technologies for sustainability [1].

AI technologies are powerful development accelerators because of their ability to optimize resource allocation and improve predictive modeling and real-time monitoring for evidence-based decision-making, thus increasing the efficiency of complex socio-technical systems. In sustainability, these are significant positive impacts across climate, health, energy, and the agricultural sectors. On the contrary, a large-scale AI lack of transparency, access, and inequity, algorithmic biases, and energy consumption is AI sustainability paradoxes [2, 3].

This chapter reviews current literature focusing on the role of AI technology and improving sustainability from 2020 to 2023. Out of these, the specific objectives are to: (1) analyze the application of AI technologies in critical areas for sustainability; (2) outline the challenges in methods and theory, and in ethics and governance; (3) examine new avenues for the challenges of policy and governance. This evidence-based, peer-reviewed research, along with systematic and high-quality conference papers, provide the most equitable and positive outlook on sustainability and the development of AI technologies.

2 Conceptual Foundations: AI and the SDGs

The relationships between and among AI technology and the U.N. SDGs have been investigated in the past. The first to map the way AI technologies affect/both positively and negatively the movement of target attainment of SDGs was Vinuesa et al. [1]. This conceptual framework, while focused on SDG, provides a way to locate sector-specific AI applications in the relevant domains of social, economic, and environmental development (Energy, Health Care, Agriculture).

In practice, AI serves as an aid in achieving the SDGs through advanced, data-informed decision-making and system enhancement. Tools like climate change models, healthcare diagnostics, and natural disaster prediction amplify response and planning capabilities. Also, in transportation, logistics, and energy systems, resource consumption is lowered, and automation and system efficiencies are improved. These features facilitate the supportive evidence and a based intervention system cross-sectionally in different areas and are sustainability features aligned [4, 5].

The increasing footprint in the environment for sizable AI models has raised concerns about the potential impacts of bias, opacity, and lack of control on the world, which has an outlying urgent necessary call. Within the frameworks seeking to balance ethics and controls, there is increasing emphasis on inclusive, accountable, and transparent designs to ensure that the AI systems enhancing capability toward SDGs in poorer areas do not deepen inequity and erosion of the environment. These are the convictions that must be adhered to for the improvement of AI systems to impact positively on the SDGs further [3, 6].

3 Climate Change: Prediction, Mitigation, and Adaptation

There are several realms of climate science. The role of artificial intelligence is significant and growing in climate modeling, planning for policy decisions, climate emissions reduction, adaptation, and more. Nonlinear relationships and interactions among Earth's systems are encapsulated in climate models, whose resolution and precision are improved by the application of machine learning. Research, such as survey by Rolnick et al. [5], encased a plethora of ML applicability in solving climate change problems and challenges, illustrating and emphasizing instances of optimization of energy systems and efficient processes in industries, monitoring emissions of all scales etc.

Figure 1 outlines the main areas where artificial intelligence is helping with climate action: climate change modeling and prediction, monitoring and reduction of emissions, optimization of energy and industries, forecasting climate change extremes, planning and adapting to climate change risks, and developing sustainable AI systems that mitigate environmental harm.

Designating specific AI research activities that help mitigate climate change as a priority is emerging as one important approach. According to Kaack et al. [4], some of the most important activities necessary for climate change mitigation include focusing on the most emission-heavy applications of AI, as well as ensuring that

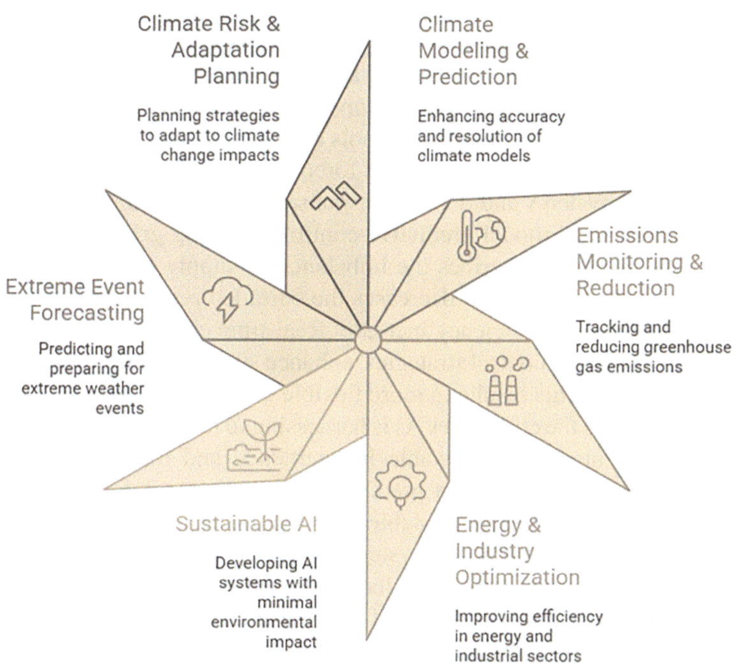

Fig. 1 AI's role in climate action across mitigation, adaptation, and sustainability domains

the carbon emissions of the AI models are minimized. Incorporating practices for the efficient training of models, the selection of climate-relevant tasks, and the use of sustainability as a criterion for evaluating AI research are ways of ensuring that AI serves the climate change mitigation actions positively rather than encumbering the efforts with initiatives that focus solely on the problems and leave no actionable climate-emissions-reducing activities.

Recent innovations in the development of AI models have ushered in a new era in the use of AI in climate change forecasting and mitigation. It is now possible to use transformers and attention mechanisms to model long-term climate time series with great accuracy and to accurately model complex spatio-temporal relationships, thus improving the forecasting of extreme climate events and climate variability [7]. AI tools for climate risk assessments also enhance the effectiveness of climate adaptation by pinpointing locations at risk from climate-related events and proposing actions to improve capacity to cope with these events and reduce the risks associated with climate change [8].

4 Energy Systems, Smart Grids, and Renewable Integration

Modern power grids face challenges related to intermittency, forecasting, and real-time control due to the integration of variable renewable energy sources such as solar and wind. Specific short-term forecasting, adaptive load management, and real-time control are required in renewable energy forecasting. Zhang et al. [9, 10] mentioned that machine learning and deep learning tools are effective in improving the stability and reliability of the grid through demand forecasting and real-time energy control.

Energy storage systems and demand response system control are based on reinforcement learning and model predictive control, allowing grid operators to learn from their systems. This describes the imbalance of supply and demand in uncertain and variable systems, and it decreases the curtailed renewable energy. Overall system performance and efficiency increase. Real-time changes in grid conditions, combined with AI-based control strategies, enhance system control, adaptability, and overall performance. This results in more flexible energy systems.

Figure 2 addresses the challenges AI is harnessing to tackle smart grid difficulties, particularly in the areas of Renewable Intermittency and forecasting uncertainty, through AI-approached and driven predictive forecasting, reinforcement learning-based optimization, and secure, scalable grid architectures toward accomplishing stable, efficient, and resilient power systems.

The advent of privacy-preserving federated learning and AI-powered digital twin simulations boosts scalability and trust of smart grid operations. Digital twins create virtual copies of energy assets for simulation, testing, and optimization, while federated learning streamlines collaborative optimization of distributed energy resources, even when sensitive consumption and operational data remain in isolated silos.

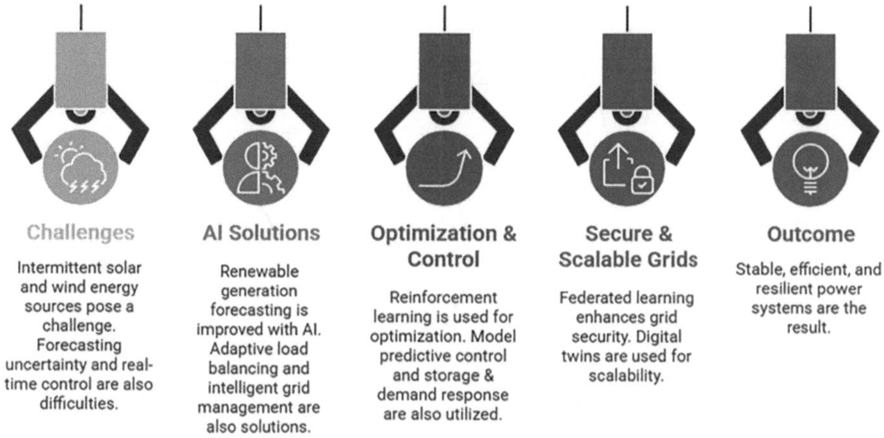

Fig. 2 AI-enabled smart grids for renewable energy integration and resilience

These techniques enable privacy-preserving and data-informed seamless integration of renewables across distributed energy systems [11, 12].

5 Agriculture and Food Security

The implementation of artificial intelligence in the automation of precision farming has improved productivity and sustainability levels within the agricultural sector. Enhanced monitoring of crops, better resource optimization, disease detection, and yield forecasting allows the agricultural sector to diminish resource inefficiencies. In cooperation with sensor, drone, and satellite data, farmers can make timely, precise decisions to target SDG 2 (Zero Hunger) and support the efficient production of food [13].

Another step forward is privacy-preserving collaboration in AI-driven agriculture. Federated learning enables various farms and agricultural institutions to collaborate in training models for disease detection and decision-support systems without exchanging sensitive information. This model allows knowledge transfer across different regions of the world while respecting local data ownership and privacy. It is, therefore, ideal in distributed and smallholder farming ecosystems [14].

Figure 3 demonstrates the ways through which AI facilitates data-driven collaborative systems for intelligent privacy in the optimization of agriculture to achieve resource efficiency, improve productivity, and promote inclusivity alongside energy-efficient agricultural systems in line with the SDG 2 (Zero Hunger) framework.

Equity and sustainability are also proposed in recent studies focusing on the development of lightweight and energy-efficient AI models. These models can be deployed on edge devices within resource-constrained farm environments. This lessens the dependence on cloud infrastructure for data processing, decreases the overall energy

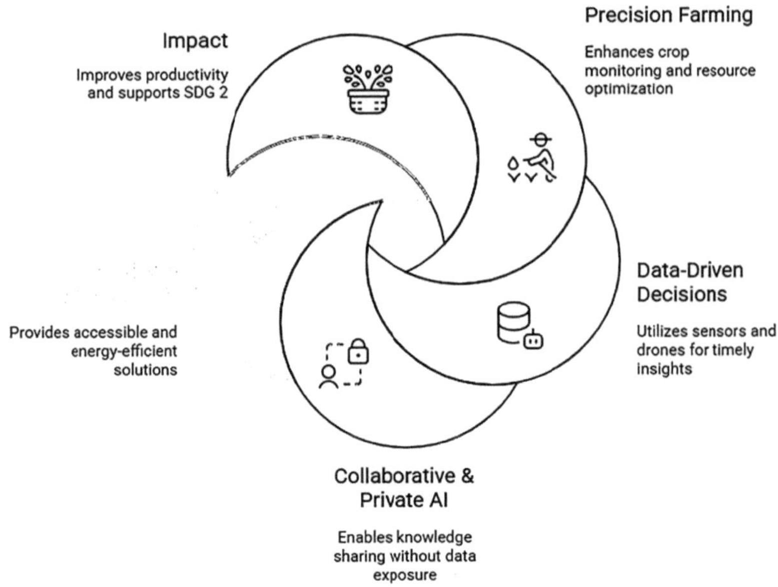

Fig. 3 Role of AI in agriculture for sustainable and inclusive food systems

footprint, and democratizes access to advanced AI technologies in rural and low-income populations. This is critical to the inclusive and environmentally sustainable scaling of AI-driven agriculture [15].

6 Healthcare and Equitable Access

The advancement of artificial intelligence in diagnosis and treatment can positively transform and offer accessible care in low resource and underserved geographical areas. It can also improve diagnostic and clinical triage workflows form digital domains to optimize clinical decision-support systems. AI's integration in telemedicine allows real-time remote patient consultations and ongoing patient surveillance, which is beneficial to remote and low income underserved clinical locations [16, 17].

Nonetheless, there are several aspects regarding the consequences and impact of artificial intelligence support systems in healthcare. Clinical decision systems are especially vulnerable when non-transparent systems are used, as they may create a form of disengagement and lose accountability to the healthcare systems. Non transparent systems may also foster biases to the extent that systems may learn low dimensional aspects of the target set and neglect unlearned aspects. Consequently, healthcare systems have concentrated adjusting the use of artificial intelligence systems to mitigate the consequences bias [18].

7 Smart Cities, Transportation, and Urban Resilience

The importance of AI in cities includes, but is not limited to, adaptive control of traffic systems, surveillance and control of air and noise pollutants, management of energy, urban planning, and other actions. One example of such applications is adaptive controllers for traffic signal control systems. Systems that employ AI and ML technologies have been reported to alleviate system congestion, fuel savings, and even greenhouse gas (GHG) reduction [19].

For urban planners, the absence of traffic management is compensated for by digitally twinned cities and AI-powered simulation platforms that allow for integrated predictive planning. These platforms allow planners to design urban systems, evaluate infrastructure investment strategically, and simulate the impact of policies on development for several scenarios over time. These urban AI systems, in the presence of good governance, ethical planning frameworks, and AI for urban systems, have the potential to equitably solve the urban systems challenges and build trust in smart city collaborations [20, 21].

Research on transport AI prioritizes Adaptive Multi-Objective Optimization which considers efficiency, environment, and social equity. Modern AIs integrate more variables, as opposed to solely focusing on minimizing travel time and are able to optimize on emission reduction, accessibility, and equitability by demographics. Equity in optimization is critical to developing systems and to the transport systems to integrate and support the resilience and sustainability of contemporary cities and their systems [22].

8 The Use of Artificial Intelligence in the Monitoring of the Environment, the Biodiversity, and the Oceans

Advanced machine learning enables large-scale rapid monitoring of the environment which can be used to trigger conservation actions. Automated systems for classifying land use, mapping habitats, and monitoring biodiversity can be used in large remote areas by integrating remote sensing with Convolutional Neural Networks (CNNs). These systems improve the efficiency of monitoring the environment and trigger the rapid response needed for deforestation, significant ecological changes, and species monitoring [23].

AI helps researchers track the pollution of the ocean and assess the overall health of the ocean. Researchers use satellite and satellite technology to track and analyze the quality and biodiversity of the water. AI helps edge systems monitor and reduce industrial pollution. These systems help fulfill the UN Sustainable Development Goals, especially SDG 14 (Life Below Water) and SDG 15 (Life on Land) because they provide solutions to technical challenges involving environmental protection and the sustainable use of economic resources [24, 25].

9 Circular Economy, Waste Management, and Resource Optimization

The impact of artificial intelligence (AI) and its ability to foster circular economy principles is being appreciated to a greater extent. The recognition of attained efficiency in different stages of a product's life cycle, in supply chains, and in waste management continues to grow. Organizations achieve reductions in overproduction and waste and increase resource efficiency through AI-driven demand forecasting, inventory management, and product life cycle analytics. Waste predictive models help assess and pattern waste in the various flows of materials (i.e., material flows) and plan for the opportunities to reuse, recycle, and recover [26, 27].

.

Figure 4 illustrates the ways through which AI sustains circular economy processes by efficient supply chain optimization, active forecasting of waste streams, waste management, and resource recovery, ultimately increasing cross-section emissions reduction and sustainability impact throughout the product life cycle.

Specific to waste management, AI methods, including computer vision, robotics, and optimization algorithms, are used to automate waste sorting and enhance collection logistics. Gupta et al. [28] showed that AI waste-sorting systems improve material recovery, while intelligent routing algorithms enhance collection schedules and vehicle routing. These methods improve operational cost efficiency and decrease greenhouse gas emissions by curbing fuel use, lessening landfill use, and making urban ecosystems more sustainable and resource-efficient.

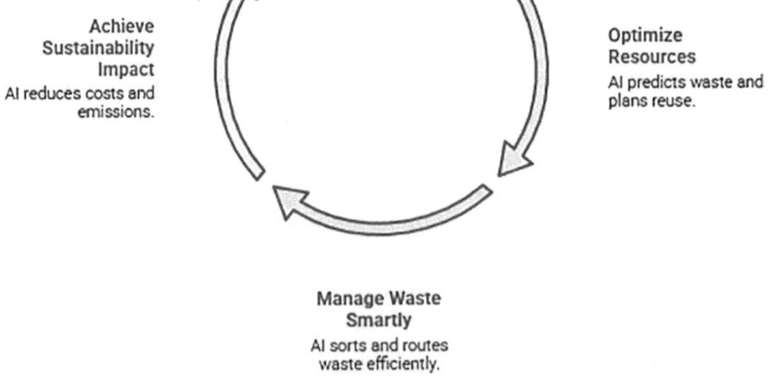

Fig. 4 AI-driven circular economy cycle for sustainable resource management

10 Methodological Advances: Green AI, Federated Learning, and Digital Twins

10.1 Green AI

Green and energy-efficient AI concerns itself with how the field of AI and its applications impact the environment, and how to mitigate the impact of AI systems on the climate. AI applications involve the training and deployment of models, and the models, when in use, are resource-heavy when it comes to processing. Large AI models consume a lot of energy and even generate carbon emissions. It will be useful in the future to focus on building lighter models, and building efficient learning algorithms that cut down on resource use and improve performance on resource usage. Enhancing the performance of algorithms on resource usage will be of great importance in climate applications that are deployed in resource-constrained settings. Such settings include edge devices, rural areas, and developing countries.

Green AI focuses on deploying models and training strategies that improve energy efficiency and building systems to focus on neural architecture search (NAS), quantization, and pruning to improve energy efficiency [29, 30].

10.2 Federated and Privacy-Preserving Learning

The development of federated and privacy-preserving learning as a form of dual-use technology is a key enabler of collaborative AI. Federated learning allows participants (e.g., institutions conducting research, governmental organizations, and healthcare corporations) to collaboratively train machine learning (ML) models without exposing themselves to privacy breaches as they do not share raw data. They only share weights after every iteration of model training. This unique form of model training also addresses regulatory requirements for data ownership and sovereignty as privacy risks are significantly reduced. This form of model training is less relevant to sustainability and other applications where data is sparse, heterogeneous, distributed, and sensitive.

Analytics across healthcare and climate modeling collaborations are other examples of integrating federated learning without the foremost centralization of a data repository; they allow cross-region and cross-institution collaborations to occur at greater scales. Partial data privacy is compromised for the other local participants, and for federated learning to facilitate the development of reliable AI, privacy must be preserved to a great extent. Such privacy is further enhanced by privacy-preserving techniques such as secure aggregation and differential privacy. Such federated methods do not only shield privacy but also offer a fair share of the model performance and aggregated Success [12, 31].

10.3 Digital Twins

Digital twins powered by AI are architectural and engineering marvels that support real-time modeling and optimization of physical systems. Digital twins create virtual replicas of bodily systems, enabling continuous monitoring and prediction. With the aid of AI and the digital twin, virtual models determine efficiency and efficacy by predicting possible outcomes using datasets, adapting systems on the fly, and driving proactive decision-making developed for optimal efficiency and sustainability.

Digital twins are now being used for renewable energy farms, infrastructure, and cities. Stakeholders can conduct virtual scenario changes to eliminate or reduce the risks of real-world implementations. Digital twins eliminate the costly trial-and-error approach by integrating technology and enabling informed decision-making for sustainability [11, 20].

11 Governance, Ethics, and Social Implications

Trust and fairness in the positive and long-term impact of the innovations and applications of the AI systems for sustainability will need ethical conduct for their development and execution. AI ethics need to be framed around the concepts of fairness and transparency and accountability [3, 6]. In the context of sustainability, unreasoned and unfair AI intervention biases would have an impact on the disadvantaged groups and the unequal distribution of environmental concerns and access to ecological resources. Therefore, the AI systems in the particular domains are being strengthened for inclusive and responsible action [32].

Apart from the above, the Effects of Ethics in the governance of Society, the sustainability of AI Impacts, and the Sustainability of the AI Impacts of deep sustainability initiatives for Specific Domains are sustained by the governance of society and such impacts. Integration of policy and multi-stakeholder governance models provides for the development of incorporation to address climate change, its impacts on social and economic development, biological diversity, conservation tools, and resource management at the global level [6, 31].

12 Emerging Research Directions and Open Challenges

In the next five years, the various roles of AI in fostering sustainability will be shaped by diverse emerging research directions. The development of energy-efficient foundation models that achieve a desirable trade-off between performance and reductions in computational and carbon footprints will be one of the focus areas. Of particular interest will be carbon accounting, energy-optimized training routines, and sustainable hardware-software co-design frameworks [29, 30]. Another focus area will be federated climate science, comprising the use of federated learning for collaborative climate modeling to avoid the centralization of critical and/or sensitive data, thereby enabling large-scale collaborations with data-empowered climate modeling and data sovereignty [31].

The third area concerns policy and governance as it pertains to AI systems that simplify or "translate" advancements from complex models into forms that are actionable and relatable to policymakers. From this angle, explainable, human-centered AI has the potential to enhance the connection between predictive insights and policy-related decisions based on booking status, increasing trust, accountability, and evidence-driven governance [18]. In the same vein, AI that optimizes the circular economy is becoming really popular and concerns the application of AI throughout the life cycle of a product from design and manufacture to use, recycling, and disposal so that waste is minimized, resource use is optimized, and sustainable production and consumption patterns are promoted [26].

With all of this insight, there are still some unsettled issues. These concerns the alignment of economic and institutional incentives with sustainability frameworks, the design, and implementation of standard lifecycle assessment protocols for AI systems and the unequal distribution of AI-driven tools for sustainability and their equitable use. Solving these issues will depend on the collaboration of multiple disciplines, the creation of standard datasets and markers to ensure reproducibility and fairness, and the joint action of scholars, policymakers, and industries to deliver AI solutions for sustainability that are fair, unambiguous, and beneficial for all.

13 Policy Recommendations

To maximize the usefulness of AI for sustainability purposes, specific and actionable policies will need to be put in place. First of all, public funding agencies and governments should require carbon accounting for training machine learning models and for inference, especially for publicly funded research. Standardized energy and carbon consumption tracking will be useful to promote transparency and adoption of more energy-efficient models.

Second, policymakers need to support federated and decentralized AI systems for secure, privacy-preserving, yet collaborative work across institutions in sensitive areas like climate science, healthcare, and agriculture. Such systems enable large-scale collaborative efforts, foster trust, and accelerate innovation. Also, significant funding for open, federated, and decentralized datasets and evaluation benchmarks for sustainability will be necessary to ensure better reproducibility, equity, and improved comparability of AI for sustainability research.

Institutions and frameworks need to be put in place to ensure that fairness, transparency, and accountability are communicated to the public regarding deployed AI systems and that decisions affecting people and communities are ethical and justifiable. Policies must also be directed toward establishing educational and computational resource facilities in the lower- and middle-income countries so that the inequalities created by AI are reduced, and the sustainable development of AI systems is equitably distributed worldwide.

14 Conclusion

AI is advancing rapidly and optimistically, becoming a primary contributor to sustainable development by enhancing decision-making, resource use, and resilience. 2020–2025 has seen the implementation of AI in various fields to support and sustain developments. Sustainability in climate and energy systems, and in the efficient use of other fields such as agriculture, health care, urban development, and conservation, as well as planning, is critical. This body of work also acknowledges that AI in its design and deployment, must take into account the so-called work, energy, and ethical frameworks that encompass governance to mitigate the risks of bias and inequality, unintentional energy consumption, and ethical frameworks.

Over the horizon, the anticipated positive impact of AI on the Sustainable Development Goals will depend on the further development of responsible and inclusive AI practices. Research and policy priorities include developing energy-efficient, low-carbon AI models, broadening the adoption of federated and privacy-preserving collaborations, enhancing explainability to enable evidence-based policy formulation, and equitably allocating AI across regions and communities. AI has the potential to be positive and to become a powerful instrument to facilitate ethically balanced, innovative technological development, as well as cohesive and sustainable development in the future.

References

1. Vinuesa, R., Azizpour, H., Leite, I., Balaam, M., Dignum, V., Domisch, S., Felländer, A., Langhans, S. D., Tegmark, M., & Nerini, F. F. (2020). The role of artificial intelligence in achieving the Sustainable Development Goals. *Nature Communications, 11*(1), 233. https://doi.org/10.1038/s41467-019-14108-y
2. Cowls, J., Tsamados, A., Taddeo, M., & Floridi, L. (2021). AI and sustainability: Opportunities and challenges. *AI & Society, 36*(3), 703–715. https://doi.org/10.1007/s00146-021-01294-x
3. Patel, V., Desai, M., & Rao, A. (2024). Ethical AI frameworks for sustainable development: Principles and practices. *Journal of Business Ethics, 178*(2), 301–317. https://doi.org/10.1007/s10551-023-05448-3
4. Kaack, L. H., Donti, P. L., Strubell, E., Kamiya, G., Creutzig, F., & Rolnick, D. (2022). Aligning artificial intelligence with climate change mitigation. *Nature Climate Change, 12*, 518–527. https://doi.org/10.1038/s41558-022-01377-7
5. Rolnick, D., Donti, P. L., Kaack, L. H., Kochanski, K., Lacoste, A., Sankaran, K., Ross, A. S., Milojevic-Dupont, N., Jaques, N., & Waldman-Brown, A. (2022). Tackling climate change with machine learning. *ACM Computing Surveys, 55*(2), 1–96. https://doi.org/10.1145/3485128
6. Ramos, C., Torres, A., & Mendoza, L. (2024). AI governance for sustainable development goals: A multi-stakeholder framework. *Global Environmental Change, 84*, 102118. https://doi.org/10.1016/j.gloenvcha.2023.102118

7. Qiao, S., Zhang, T., & Liu, Y. (2023). Transformer models for climate prediction: A comparative study. *Geoscientific Model Development, 16*(8), 2456–2472. https://doi.org/10.5194/gmd-16-2456-2023
8. Li, W., Chen, Y., Zhang, T., & Liu, J. (2023). AI for climate risk assessment and adaptation: A global perspective. *Nature Sustainability, 6*(5), 512–525. https://doi.org/10.1038/s41893-023-01088-7
9. Zhang, Q., Li, H., Wang, Y., & Chen, W. (2023a). AI-driven smart grids for renewable energy integration: A comprehensive review. *Renewable and Sustainable Energy Reviews, 177*, 113125. https://doi.org/10.1016/j.rser.2023.113125
10. Zhang, H., Liu, X., Wang, Y., & Zhao, Q. (2023b). AI-assisted policy simulation for sustainable development: A dynamic systems approach. *Policy Sciences, 56*(3), 401–420. https://doi.org/10.1007/s11077-023-09501-2
11. Cheng, H., Wang, X., & Liu, Z. (2024). Digital twins for renewable energy farms: Real-time monitoring and optimization. *Energy Reports, 10*, 210–225. https://doi.org/10.1016/j.egyr.2023.12.045
12. Saxena, N., Goyal, P., & Tiwari, S. (2023). Privacy-preserving AI for smart grids: A federated learning approach. *IEEE Transactions on Smart Grid, 14*(2), 1450–1465. https://doi.org/10.1109/TSG.2022.3221456
13. Sharma, M., Gupta, P., Verma, A., & Kumar, N. (2024). AI in precision agriculture: A systematic review of machine learning applications for crop management. *Computers and Electronics in Agriculture, 216*, 108122. https://doi.org/10.1016/j.compag.2023.108122
14. Nair, V., Patel, R., & Mehta, S. (2024). Federated learning for smart agriculture: Collaborative crop disease detection. *Agricultural Systems, 214*, 103118. https://doi.org/10.1016/j.agsy.2023.103118
15. Kumar, S., & Singh, R. (2024). Edge AI for environmental monitoring: A scalable and privacy-preserving approach. *Environmental Modelling & Software, 161*, 105118. https://doi.org/10.1016/j.envsoft.2023.105118
16. Biswas, S., Roy, S., & Das, A. (2023). AI for equitable healthcare in developing regions: A case study in telemedicine. *The Lancet Digital Health, 5*(6), e345–e355. https://doi.org/10.1016/S2589-7500(23)00087-2
17. Esteva, A., Chou, K., Yeung, S., Naik, N., Madani, A., Mottaghi, A., Liu, Y., Topol, E., Dean, J., & Socher, R. (2021). Deep learning–enabled medical diagnosis for sustainable healthcare. *Nature Medicine, 27*(1), 24–27. https://doi.org/10.1038/s41591-020-01198-1
18. Ghai, B., Liao, Q. V., Zhang, Y., Bellamy, R. K. E., & Mueller, K. (2023). Explainable AI for sustainable policymaking: A case study in climate adaptation. *AI & Society, 38*(4), 2105–2120. https://doi.org/10.1007/s00146-023-01660-1
19. Xu, Y., Wang, Z., Li, J., & Chen, H. (2022). Reinforcement learning for adaptive traffic signal control in smart cities. *Transportation Research Part C: Emerging Technologies, 134*, 103119. https://doi.org/10.1016/j.trc.2021.103119
20. Chen, C., Liu, X., Zhang, Y., & Zhou, M. (2023). Digital twins for sustainable urban planning: A framework and case study. *Journal of Urban Technology, 30*(2), 45–67. https://doi.org/10.1080/10630732.2023.2189456
21. Winkler, T., Schmidt, A., & Müller, J. (2024). AI in smart cities: A review of applications and governance frameworks. *Sustainable Cities and Society, 90*, 104119. https://doi.org/10.1016/j.scs.2023.104119
22. Patil, S., Nair, V., & Deshmukh, A. (2023). AI in sustainable transportation: A bibliometric review and future research agenda. *Transportation Research Part D: Transport and Environment, 115*, 103119. https://doi.org/10.1016/j.trd.2023.103119
23. Rahman, M., Islam, S., & Hossain, M. (2024). AI-driven biodiversity monitoring using satellite imagery and convolutional neural networks. *Remote Sensing of Environment, 300*, 113128. https://doi.org/10.1016/j.rse.2023.113128
24. Park, J., Kim, S., & Lee, H. (2024). AI for ocean conservation and monitoring: Detecting marine pollution with deep learning. *Marine Policy, 159*, 105118. https://doi.org/10.1016/j.marpol.2023.105118

25. Silva, L., Costa, M., & Almeida, J. (2023). Edge AI for real-time pollution control in industrial zones. *Environmental Monitoring and Assessment, 195*(7), 845. https://doi.org/10.1007/s10661-023-11302-8
26. Fernandez, P., López, M., García, S., & Martinez, R. (2023). AI and circular economy: A systematic review of applications in waste reduction and resource optimization. *Journal of Cleaner Production, 382*, 135150. https://doi.org/10.1016/j.jclepro.2022.135150
27. Lopez, M., Singh, A., & Chen, Y. (2024). AI for sustainable supply chain optimization: A multi-objective approach. *International Journal of Production Economics, 268*, 108122. https://doi.org/10.1016/j.ijpe.2023.108122
28. Gupta, A., Reddy, S., & Mehta, P. (2024). AI in waste management: Trends, challenges, and future directions. *Waste Management, 174*, 210–225. https://doi.org/10.1016/j.wasman.2023.12.010
29. Nguyen, T., Wang, J., Chen, Y., & Zhang, L. (2023). Green AI: Energy-efficient deep learning models through neural architecture search and quantization. In *Proceedings of the AAAI Conference on Artificial Intelligence, 37*(8), 9456–9464. https://doi.org/10.1609/aaai.v37i8.26145
30. Zhu, Z., Xie, X., & Tan, Y. (2024). Neural architecture search for green AI models: Balancing accuracy and energy efficiency. *IEEE Transactions on Neural Networks and Learning Systems, 35*(3), 1120–1135. https://doi.org/10.1109/TNNLS.2023.3312456
31. Wang, L., Zhang, H., Kim, S., & Tanaka, K. (2023). Federated learning for cross-border climate modeling: Enhancing global cooperation. *Climate Informatics, 12*, 88–102. https://doi.org/10.1002/cli2.12045
32. Gomez, R., Silva, L., & Fernandez, J. (2024). Fairness-aware AI for environmental justice: Mitigating bias in climate data. *Environmental Science & Technology, 58*(12), 5201–5215. https://doi.org/10.1021/acs.est.3c09845

AI-Driven Quality Inspection for Sustainable Manufacturing in Industry 4.0

K. Indra Gandhi, P. K. Jawahar, G. Kannan, Abdul Rajak A. R., and Ashutosh Mishra

Abstract In assembly lines and in logistics operation modern connections, there are defective items that are especially those products whose components are either absent or wrongly fitted and this presents a great challenge. Such defects have the potential to hamper the working process, raise the cost of production, and adversely affect the satisfaction of customers. These mistakes may pass unobserved at some point in the production process like assembly as well as packaging since they are done either through human or robotic supervision. To solve this problem, this paper proposes an automated inspection system powered by AI, which will inspect the missing components in a very precise way using computer vision. The offered methodology is based on the use of transfer learning, experimenting with a lightweight MobileNet Convolutional Neural Network that was retrained using a collection of some 20,000 labeled pictures. Classification was between 32 and 98 percent accurate, and the model is demonstrably cost-effective as compared to traditional manual inspection tasks. In line with the concepts of Industry 4.0, the system reflects the edge-to-cloud scheme, with the data of inspection, such as product ID, number of missing components, sent to the ThingSpeak cloud service to be monitored and analyzed remotely. The cloud

The original version of the chapter has been revised. A correction to this chapter can be found at https://doi.org/10.1007/978-3-032-28217-0_12

K. Indra Gandhi (✉) · P. K. Jawahar · G. Kannan
Department of ECE, B.S. Abdur Rahman Crescent Institute of Science & Technology, Chennai, India
e-mail: indra@crescent.education

P. K. Jawahar
e-mail: jawahar@crescent.education

G. Kannan
e-mail: kannan@crescent.education

A. Rajak A. R. · A. Mishra
Birla Institute of Technology and Science, Pilani, Dubai, United Arab Emirates
e-mail: abdulrazak@dubai.bits-pilani.ac.in

A. Mishra
e-mail: ashutosh@dubai.bits-pilani.ac.in

A. Khanna et al. (eds.), *Artificial Intelligence for Sustainable Development Goals*,
https://doi.org/10.1007/978-3-032-28217-0_5

dashboard is user friendly and has visual intelligence setting trending results and also indicates products with more than 20 component shortages. Also, product images are posted to be verified and tracked and this makes it a transparent and efficient quality assurance procedure. The advantage of such an integration of edge computing and cloud based analytics is that it allows decisions to be taken in real time and helps to gain better sustainability, operational efficiency, and reliability of products in smart manufacturing.

Keywords Industry 4.0 · Logistics · Computer vision · MobileNet · Real-time model

1 Introduction

Industry 4.0 is transforming manufacturing, optimization, and distribution of goods by companies. They are being transformed to smarter manufacturing by integrating smart technologies within the manufacturing system like the Internet of Things (IoT), cloud, analytics and artificial intelligence (AI) and machine learning. Particularly, the manufacturing sector is bound to gain a lot by this revolution in the production of smart factories. Such intelligent systems help utilize loads of data that the sensors at the facility can capture on-site to provide real-time performance insight on equipment and allow predictive maintenance, thereby limiting downtimes and unexpected outages. The overview of the major technologies contributing to the realization of the Industry 4.0 is provided in Fig. 1 [2, 3]. Besides, the inclusion of automation also contributes better to the company ability to predict and avoid failure of equipment effectively. It allows identifying any possible issue in the performance of the equipment and getting rid of it with the help provided by artificial intelligence (AI)—informed analytics monitoring equipment round-the-clock and making sure that no give faults are going to cause downtime. Techniques of object detection form an essential part of artificial intelligence in that they identify and position certain objects in an image plucked out of video. Such object detection methods as HOG, SSD, YOLO, Deep SORT, etc., are implemented in major spheres of logistics, including Automated Inventory Management, Pallet and Package Detection, Quality Control and Inspection, Order Picking and Sorting, Real-time Tracking and Tracing, and Yard Management [1].

Artificial intelligence (AI) involves computer vision, which permits visual information such as images and videos to be interpreted and valuable information to be extracted by computers. It is based on human visual system and involves the use of cameras, advanced algorithms, and huge datasets, etc., that it uses to perform visual analysis that often surpasses humans in speed and accuracy [4]. With the help of the power of computer vision, the systems will be able to process different visual information and find defects and make the decisions, which a human observer could not notice. This technology has become increasingly essential in sectors like manufacturing and quality assurance. Its capacity to analyze hundreds of products

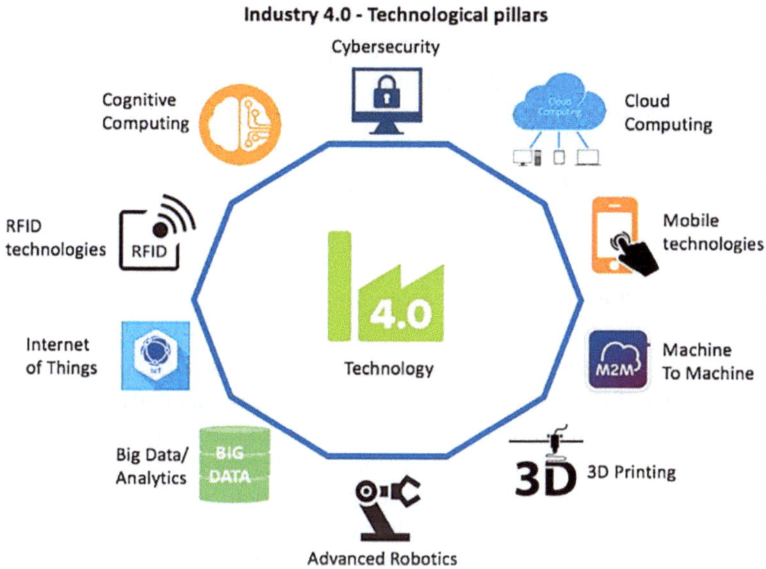

Fig. 1 Industry 4.0 and its pillars. *Source* Saturno et al. [1]

or processes per minute enables it to outperform human inspectors, particularly in identifying subtle defects [5]. Automated inspection systems that integrate computer vision enhance operational efficiency, lower production costs, and boost product quality by minimizing human errors and the fatigue typically associated with repetitive visual inspections [3]. Computer vision enables continuous, real-time comprehension of processes. However, when visual data is combined with information from machinery and other non-visual operator inputs, the output from computer vision systems becomes significantly more valuable. For seamless operation, the system should be capable of quickly sending signals to interact and communicate with human operators. Achieving this requires a computer vision framework that aligns with Industry 4.0 principles. Convolutional Neural Networks (CNNs) such as Inception, ResNet, and MobileNet are widely utilized for image classification tasks [6]. Among them, MobileNet stands out as an efficient deep learning model designed specifically for computer vision tasks on resource-constrained devices like mobile phones and embedded systems. Traditionally, machine learning and deep learning models were built separately, each optimized for a specific application and required complete retraining when the underlying data distribution changed. Transfer learning already presents a new form of solving the problem by enabling the use of previously trained models to be modified to new yet similar tasks. This strategy requires fewer models to be trained, and development cycles are reduced, as well as the increased flexibility of solutions to different areas.

The given paper is devoted to the use of computer vision in the industrial sphere, to be more exact, it concerns the detection of the lack of elements in the process of manufacturing. The assembly line was easily integrated with a robust system used to check component presence, based on computer vision algorithms and transfer learning-based models, and it did not affect usual production routines in any way. The system can be able to achieve this by exploiting the potential of computer vision, which helps the system enhance the efficacy and accuracy of missing component identification, hence, enhance product quality and lower production errors.

2 Related Work

On the one hand, it is possible to discuss the findings of Marengoni et al. [4] which have conducted an informative study into the background and advanced techniques behind the detection and classification of objects in the field of computer vision. The extensive analysis of the visual clues and other characteristics identification methods, presented in their research, gives a wide scope of various approaches to the process of related methods of detection. These characteristics are major pointers in locating, classifying objects in an image. The authors compare the efficiency of the traditional and the method of modern approach in different settings. Also, the paper raises critical review about the quality of the techniques in terms of strengths and weaknesses giving an equal view of use of a technique over another. This discussion is useful to those who deal with research and need to get familiar with the main approaches and the most recent shortcomings that question the existence of object identification systems. Sandler et al. [1] present the application of a popular deep learning model Single Shot Detector (SSD) in real-time object detection. Their primary goal is creating a fast, precise web-based system that could identify and track several classes of objects in real-time videos. In this connection, they train their model to recognize 21 object classes, such as people, animals, cars, and other objects people have every day, e.g., chairs and bottles. Their work has proven reasonably that the SSD framework is consistent in its performance in the speed of the detection, as well as accuracy in classification. The authors emphasize the usefulness of the model in practical settings especially where efficiency in computation and fast inference are of paramount importance.

In the article by Szegedy et al. [6], a detailed discussion of the history of deep learning-based object detection frameworks is presented. Their paper examines the recent development of Convolutional Neural Networks (CNNs) and related structures that have led to a massive increase in the possibility to derive high-level semantic content of visual information. It is clear that such innovations have resulted in the development of potent models that not only perform well when it comes to detection of objects but they also are robust even when set within different parameters. The essential role of their work is to help the researchers and developers define the milestones in the detection technology and offer the directions in which the technology can be improved in the future. The given paper serves as a source of positioning written as a strategic resource to guide not only research but also implementation in industry. Sehgal et al. [7] introduce comparative assessment of six object detection models, to be precise, those that detect vehicles in unstructured and quite showy road conditions. Their experiment shows subtle insights on trade-off performance of various architectures. As an example, the Faster RCNN alias ResNet 50 backbone provides a reasonable trade-off between detection performance and filed speed, so that could be applied to the situations where two aspects are important. SSD MobileNetV1, on the other hand, is much faster approximately by 31.89 percent and the accuracy suffers a little, only by a few percent points. Faster RCNN ResNet101 on the other hand has a 34.38% improvement in accuracy over SSD MobileNetV1 but also consumes more computations. The comparative research has helped give useful insights on the selection of suitable models in instances like intelligent transportation systems where speed and precision are very important with respect to the environment of application. Modern manufacturing has accepted automation; however most quality inspection systems remain relegated to manual inspection, or being based on rigid technologies that cannot cope with product appearance or assembly changes. This may commonly result in unnoticed flaws, consist of incomplete or improperly placed parts that may interrupt the manufacturing process and decrease customer satisfaction. It is evident that a more intelligent, flexible solution is required that would be able to promptly identify such problems in real-time and provide insights that could be shared without any difficulties to make wiser decisions. To fill this gap, the given work suggests the computer vision solution that can recognize missing parts with up to 98 percent of accuracy and can be done with the help of the MobileNet, transfer learning. This system has connected edge-to-cloud architecture with ThingSpeak, which facilitates real-time data to transmission and remote monitoring of the data, as well as visual dashboards. Besides the increased efficiency of the inspection process, such a move resonates with the spirit of Industry 4.0 since they provide an intelligent, easily scalable solution that can serve as a low-cost and high-level quality assurance structure.

3 Problem Statement

Most manufacturing sectors possess a great amount of parts which must be constructed into some form of a production, and the loss of a part or a misplaced part or a broken part may cause a faulty product or a manufacturing accident to take place. Computer vision can be used by the cameras or sensors to capture or record images and data of the assembly line by the cameras or the sensors, and this can be analyzed by the system to detect any instance of missing or wrong positioned components. The automatic detection enables the manufacturer to achieve the products quality, better the efficiency of the production and lessens the quantity of the products that have been defected. When visiting the Nokia Solutions and Networks Pvt. Ltd., Oragadam, Chennai the following issue on the level of assembly line was identified. Nokia Chennai factory is focused on production of the newest 5G Massive MIMO equipment, which is exported to the countries where 5G level of deployment is still highly developed. In Radio Frequency Module (RFM line) where Radio Frequency

Modules are manufactured, there are some parts called bullets and this part should be screwed in frames which will form a part of the 5G Base Transceiver Station (BTS) Antenna. Every single frame is shaped with 64 holes, which must be loaded with 64 bullets in their case. Failure to screw any of these bullets into the frame may lead to the end product being defective which may create a problem in the mind of the customers who are fortunate enough to get the defective equipment. The current paper has suggested a computer vision method based on camera sensors to capture photos or data on the assembly line. Through these inputs, the system can be able to effectively identify cases of missing or improper mounted parts.

4 Proposed Automated Training System

The operating architecture shown in Fig. 2a includes the proposed Automated Training System (ATS) that ensures sustainable and intelligent manufacturing with the United Nations, Sustainable Development Goals (SDGs), especially SDG 9 and SDG 12. In such a system, real-time video about the product frames on the assembly line is recorded by a camera. Every frame is reviewed to ensure that all the necessary parts would be correctly installed. In case the product gets the inspection, then it proceeds to the next stage of assembly. In case a missing component is detected, the system alerts the operator on an immediate basis providing visual clues to help the operator visually locate and rectify the fault.

Simultaneously, data about the critical inspection like ID of products and condition of their components is sent to the cloud (ThingSpeak) to perform advanced analysis and visualization. This guarantees the full traceability, enables on-going quality enhancement, and remote decision-making. When the correction required is made, the operator shuts off the application and the product goes ahead. This amalgamation of computer perception, electric insight, and cloud investigation will encourage accountable generation, lessen scrapping, and improve working effectiveness and as a direct consequence will lead to sustainable business innovativeness and infrastructure.

Data Collections

Figure 2b depicts the experimental setup, where the Jetson Nano, an NVIDIA heterogeneous edge computing platform, is employed. The system operates on the Ubuntu 18.04 Linux operating system. The setup comprises the Jetson Nano board connected to a 7-inch IPS touch HDMI LCD display and a 2-megapixel 1920x1080 CMOS surveillance camera. Additionally, a Bluetooth mouse and keyboard are used for user interaction and system control.

To develop the real-time dataset, video recordings were used to capture product frames at a rate of 60 frames per second, with each video lasting approximately 10–12 s, covering the entire assembly line process, as shown in Fig. 3. Two videos were recorded: one showcasing all components properly screwed in (including the bullet) and another highlighting the absence of the bullet, resulting in some components

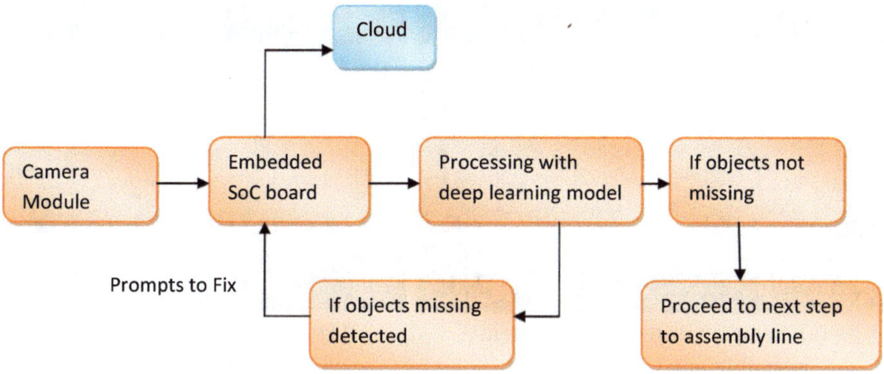

(a) Architectural design of the Proposed Automated Training System (ATS)

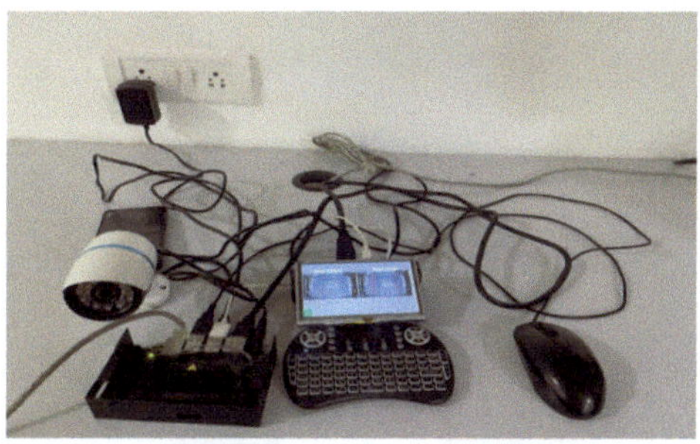

(b) Experimental Setup

Fig. 2 **a** Architectural design of the Proposed Automated Training System (ATS). **b** Experimental setup

being missing, as illustrated in Fig. 3a and b. Using Python's OpenCV library [8, 9], 300 frames were extracted from each video and organized into separate directories. A frame counter was implemented to maintain proper frame numbering. A reference frame was randomly selected and used for object annotation. Annotations were carried out to generate a dataset of images indicating the positions of objects, with the goal of improving classification accuracy. The LabelImg application was used for manual annotation, which also supports automatic class creation, as shown in Fig. 4a and b. These markups were stored in XML that could be used in training and evaluation.

Dataset

A Python-based application was implemented to convert annotation files from XML format into a compact CSV format, enabling efficient storage and retrieval of annotation data. The CSV file that is produced contains relevant data like names and sizes of the images, name of the object category to which that image belongs as well as the

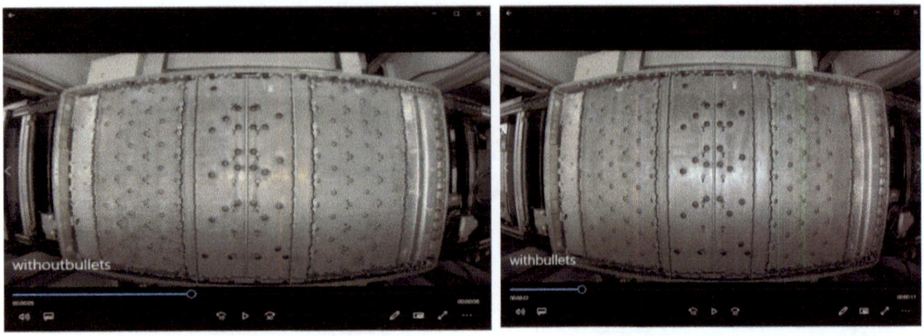

(a) Sample video of a Product in assembly line without bullet

(b) Sample video of a Product in assembly line with bullet

Fig. 3 **a** Sample video of a product in assembly line without bullet. **b** Sample video of a product in assembly line with bullet

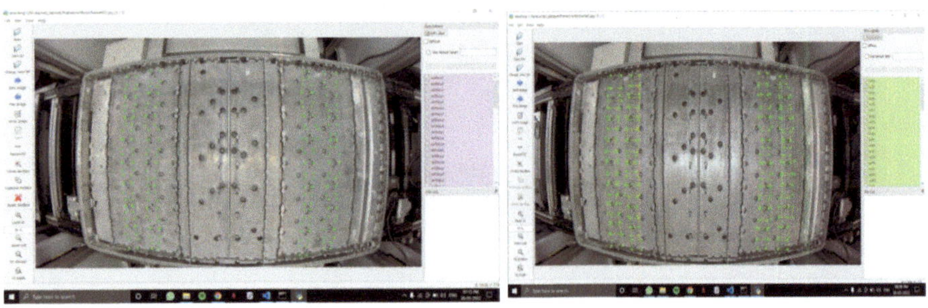

(a) Annotations made on “Without Bullets Frame”

(b) Annotations made on “With Bullets Fram

Fig. 4 **a** Annotations made on “Without Bullets Frame”. **b** Annotations made on “With Bullets Frame”.

coordinates of the bounding box. Based on this organized data, each image is subsequently cropped based on its bounding boxes to cut-off the pertinent parts. These parts that are extracted are then used to train the model and apply a more detailed examination. The CSV file created at the step above is subsequently read, and in this case, it is assumed that this file will have the bounding box or the bounding box data as relates to the images under discussion. Some of the information contained in this data include object class labels, the coordinates that define the bounding boxes, which during training and classification processes are important in the localization and extraction of the regions of interest. With the help of OpenCV, images can be loaded in a specific folder and taken out as a region of interests (ROIs), depending on the given coordinated values of the bounding box in the CSV file. The resulting ROIs are then cropped and are saved in a directory of the form as depicted in Fig. 5a and b.

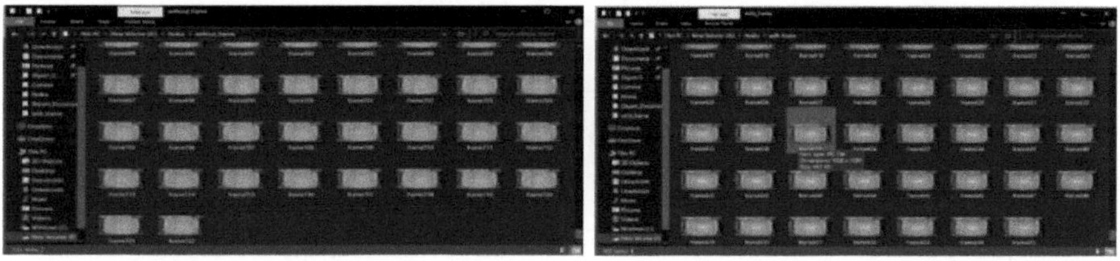

(a) Frames extracted from "Without Bullets Video"

(b) Frames extracted from "With Bullets Video"

Fig. 5 **a** Frames extracted from "Without Bullets Video". **b** Frames extracted from "With Bullets Video"

MobileNet Convolutional Neural Network

Due to the categorization into two groups of objects (with or without bullets), TensorFlow was used together with MobileNet Convolutional Neural Network (CNN). This ensemble exploited the light and potent architecture of MobileNet which is perfected to work on edge devices and was able to classify the two classes on the basis of the extracted and annotated image information. The CSV file created on the last step is read and it is assumed that it has the information about bounding boxes of objects in pictures. This data includes details such as object class labels and the coordinates defining each bounding box, which are crucial for localizing and extracting the regions of interest during the training and classification processes.

To increase accuracy without a need to train such a large dataset, the MobileNet model had been retrained with the help of transfer learning a mechanism that uses extant models to attain better results in the shortest possible time. Figure 6 illustrates a simple variant of the structure of the MobileNet-style Convolutional Neural Network (CNN) and shows the main layers aimed to perform efficient object classification work [10, 11]. First, it is required to feed an image into the network. Three dimensions are usually measured in this image; height, width, and the various channels of colors. The convolutional layer it passes through first is 32 filters and 3×3 kernel. The stride of 2 again allows to decrease the size of the picture and padding makes the final size equal. Then, it is necessary to go to the process of normalization batch, which significantly contributes to the stabilization and acceleration of the training process. It does this by modifying and normalizing the network activations, such that they will remain in the same range. It is performed per feature channel of the image and guarantees that all the layers of such network learn efficiently and are not influenced by erratic data distributions. A combination of intelligent design and training-related tricks, MobileNet is lightweight as well as powerful enough to accomplish tasks such as real-time image classification. Batch Normalization process begin by computing the mean val (mu) and variance (sigma) of each channel using the original value x i as demonstrated in Eqs. (1) and (2).

$$\mu = (1/m) * \Sigma x_i, \tag{1}$$

$$\sigma = \sqrt{\left(\frac{1}{m}\right) * \Sigma\left((x_i - \mu)^2\right)}. \quad (2)$$

The parameter m is the mini-batch size, and summation is done over the mini-batches. The second step of the process implies the normalization of the activations by each channel separately. This is achieved through the formula indicated in Eq. (3) here, but where a small positive constant epsilon is added to the denominator to avoid dividing by zero and for the numerical stability of the answer. After normalization of the activations, scaling and shifting apply in accordance of Eq. (4). The steps adjust the activations in a controlled manner and aid network training more easily

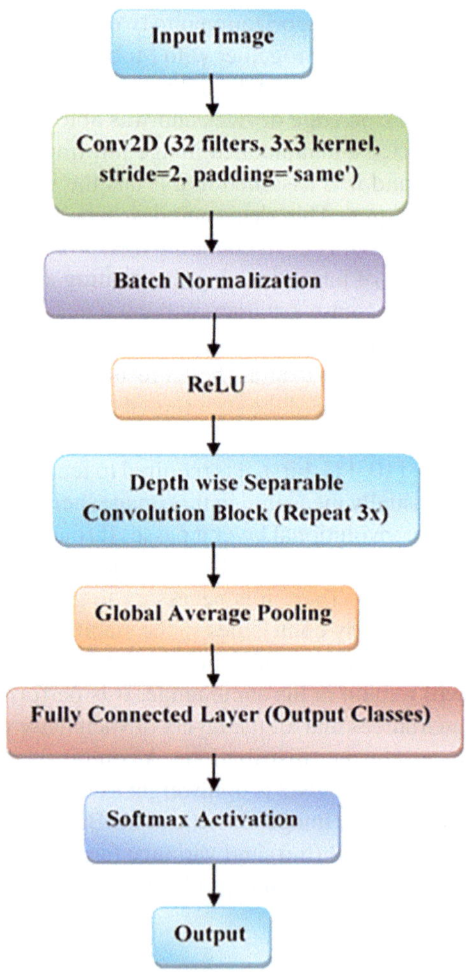

Fig. 6 Representation of a MobileNet architecture

and effectively.

$$\hat{x}_i = \frac{(x_i - \mu)}{\sqrt{\sigma^2 + \varepsilon}}, \tag{3}$$

$$y_i = \gamma^* \hat{x}_i + \beta. \tag{4}$$

γ (gamma) and β (beta) in Batch Normalization are learnable coefficients that change the scale and shift values of the normalized values $\hat{x}_i$ to, respectively, each channel, $\hat{x}_i$ is the Normalized activation and y_i is the final value to go to the next layer in the network. The Batch Normalization is used similarly, but it is used individually on each channel or feature dimension of an input tensor. In such a manner, BatchNorm can be successfully used on tensors of different shapes and dimensions, which saves the application to different neural network structure. The model can obtain better convergence and performance by normalizing the activations and adjusting with Eqs. (3) and (4) [12, 13].

Activation: The Rectified Linear Unit (ReLU) Activation is involved, under pinning to assist the model learn complex patterns. It plays the purpose of injecting nonlinearity in the network. ReLU returns the input value when it is positive and has a value of zero when it is less than zero as indicated in Eq. (5).

$$f(x) = \max(0, x). \tag{5}$$

Its derivative, given in Eq. (6), is straightforward—it is 1 when the input is greater than zero and 0 when the input is less than zero

$$f^{'(x)} = \begin{cases} 0 \text{ for } x < 0 \\ 1 \text{ for } x > 0 \end{cases}. \tag{6}$$

x may be either a number or a tensor (multi-dimensional array). This operation guarantees that the positive values are transmitted unchanged and negative values are shoved, which makes the learning faster and there is no occurrence of vanishing gradients [14].

Depth-Wise Separable Convolution Block: This particular network module is specifically useful in the lightweight neural networks, and it is composed of two steps: (1) depth-wise convolution: The input channels are processed independently (2) point-wise convolution: The outputs of the former are combined again using convolutions with size 1 × 1. The model has this structure repeated severally. Following every convolution operation, the learning is leveled out with Batch Normalization and ReLU activation is used to denote non-linearity [15, 16]. This architecture is effective and generation of cost of computing is reduced greatly and they are favorited in the resource limited area such as mobile and embedded systems.

Depth-Wise Convolution:

Input feature map size: $H \times W \times D_{\text{in}}$.
Depth-wise filter size: $F \times F$.
Depth-wise convolution output size: $H_{\text{out}} \times W_{\text{out}} \times D_{\text{in}}$.
Number of parameters in-depth-wise convolution: $F \times F \times D_{\text{in}}$.

Point-Wise Convolution: The input feature map in a Convolutional Neural Network normally possesses dimensions $H \times W \times D_{\text{in}}$ which H and W are the dimensions (spatial dimensions) and D_{in} the number of input channels. A number of 1×1 filters process this input when applying point-wise convolution. When there are D_{out}. filters, the output will be of dimensions $H \times W \times D_{\text{out}}$. The computational parameters engaged with the point-wise convolution is computed as $1 \times 1 \times D_{\text{in}} \times D_{\text{out}}$. What this implies is that all input channels are added together with a 1×1 filter to create each output channel. Conversely, depth-wise convolution processes each channel input independently, with the same number of outputs channels (D_{in}). The point-wise convolution is applied then to combines the information between the various channels and converting the depth to D_{out}. This reduces to a large extent the computational effort, and the model can learn rich feature representations.

Global Average Pooling: Pros and Cons: Global Average Pooling: Global average pooling reduces feature maps by computing average per channel thus effectively downsizing the spatial dimensions of the feature map to a single channel which is in the form of a single vector. This will aid in reducing the amount of the parameters that will keep some valuable information to be used in classification. At each channel c ($c = 1$ to C), it averages across all the H W spatial locations. This gives an output of $1 \times 1 \times C$ dimension in which each channel has the average of its respective input feature map.

$$\text{GAP}_c = \left(\frac{1}{(H \times W)}\right) * \Sigma\ (x_i, j, c). \tag{7}$$

Fully Connected Layer: The feature vector that was obtained through the global average pooling is introduced to a fully connected layer that will produce final output scores per-class so that the model can make its prediction.

Softmax Activation: The softmax is used to pull the output of the fully connected layer into probabilities on classes. To compute the probability P of a class in case of a certain input-vector $Z = [z_1, z_2, \ldots z_n]$, where n denotes the total number of classes in the given problem, the softmax computes the probability in accordance to the formula presented in Eq. (8). This guarantees that the values of any output will be in the range 0–1 and that all of them add to 1, and thus, they can be interpreted as probabilities.

$$P(\text{class } i) = \frac{\exp(z_i)}{\Sigma\left(\exp\left(z_j\right)\right)} \quad \text{for } j = 1 \text{ to } n. \tag{8}$$

Here, $\exp(z_i)$ is the exponential value of the ith component of an input-vector Z. The notation $\Sigma(\exp(z_j))$ means that we take the sum of the exponent of each of the things in the vector Z. The resulting value, $P(\text{class } i)$, will yield the predicted probability that the input will be classified in class i. This produces a positive value of each output and the overall sum of all the probabilities equals 1, which makes the output or suitable to multi-class classification. The softmax function can transform outputs to probabilities in non-negative form that sum to 1, and so is suitable when the required task is classification. In most applications, the predicted class is the one with the maximum probability obtained through argmax operation [17, 18] after passing the softmax.

Output: The model produces the outputs as probability scores of every single class. A number of critical elements are customized to favor productive training. Adam optimizer is applied to modify the weights of the model through the training process and presents an efficient and solid technique of optimization. In the case of the multi-class classification problem, the model uses categorical cross-entropy which compares the probability of the predicted estimates with the true label and then it helps the model enhance its estimates in subsequent performances. Also, accuracy is applied as a performance measure that monitors the progress of the model on how effectively it is learning the classification of inputs correctly. They combine to make the model learn effectively and test correctly with the presented data [19–21].

5 Results and Discussions

The training on the final layer of the model is started after generation of all of the intermediate feature representations usually after 4000 steps. At every stage, the system will pick a random batch of ten training images and apply the precomputed features of the images to make predictions. Comparison is then made with the actual labels after which backpropagation is used to adjust the weights of the final layer. The desired indicators of the process, e.g., training accuracy (TA), validation accuracy (VA), and cross-entropy loss (CE), are published over this process to measure the effectiveness of the learning process of the model [22–24].

Training accuracy (TA) reflects how accurately the model classifies the samples in the training set, based on the proportion of correct predictions made.

$$\text{TA} = \frac{\text{Number of Correctly Classified Samples in Trainingset}}{N}. \tag{9}$$

Validation accuracy (VA) measures how well the model performs on a separate validation dataset, which is not involved in training. It provides an unbiased assessment of the model's ability to generalize to new, unseen data.

$$\text{VA} = \frac{\text{Number of Correctly Classified Samples in Validation set}}{N}. \quad (10)$$

Cross-entropy is used as the loss function during training to measure how far the predicted probability distribution is from the actual (ground truth) distribution. The goal of training is to minimize this loss, which helps the model produce predictions that are more accurate and closely reflect the true class probabilities, ultimately improving overall performance [25].

$$\text{Cross Entropy (CE)} = -\sum_{i=1}^{N} y_i \; . \; \log(p_i), \quad (11)$$

where

$\sum_{i=1}^{N}$ indicating that we sum over all the samples in the dataset.
y_i Represents the actual label for class i (ground truth). It is **1** for the correct class and **0** for others.
p_i Refers to the predicted probability for class i, usually obtained from the softmax activation function.
N Total number of classes.

Accuracy of the Model

Figures 7a and b present the accuracy and loss trends of the retrained model throughout the training process. In Fig. 7a, the orange line represents the accuracy achieved on the training dataset, while the blue line shows accuracy on the unseen test dataset. Since the test data was not used during training, its accuracy serves as a more reliable indicator of the model's real-world performance. If training accuracy keeps increasing but test accuracy begins to decline, it suggests overfitting—a condition where the model learns the training data too well, including its noise or irrelevant patterns, and fails to generalize to new data. Monitoring both training and test accuracy helps identify and avoid such issues, ensuring the model remains robust and generalizable.

Figure 7b tracks the cross-entropy loss over time. The orange line shows how the loss decreases on the training data, while the blue line reflects the loss on the validation data. Ideally, both should decrease as the model learns. Once training is complete, the model is evaluated on a completely separate test set that was not used at any stage during training or validation. This final evaluation provides the most realistic estimate of how the model is likely to perform in real-world applications. The model's performance when tested with real-time input from the assembly line is depicted in Fig. 8a and b.

As illustrated in Fig. 8a, both training and validation accuracy show a steady upward trend when the model is tested in real-time using images from the assembly line. This consistent improvement indicates that the model is learning effectively

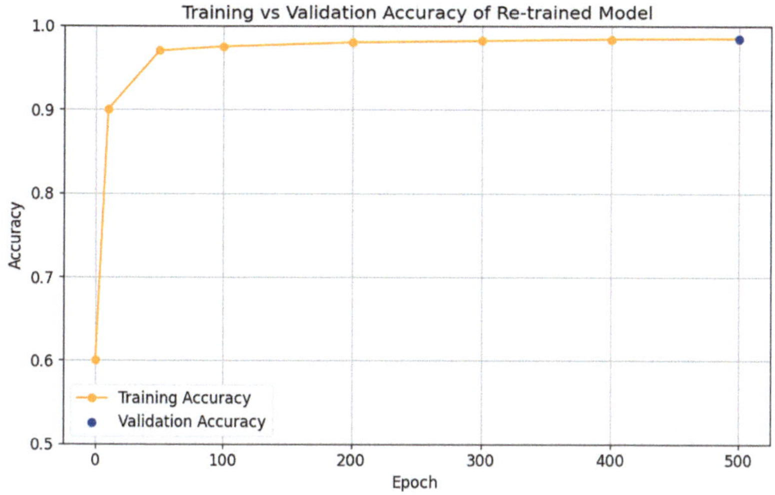

a Accuracy of the re-trained model

b Cross-entropy of the re-trained model

Fig.7 **a** Accuracy of the retrained model. **b** Cross-entropy of the retrained model

without overfitting. The close match between training and validation accuracies suggests strong generalization, meaning the model can make reliable predictions even on data it hasn't seen before an essential quality for real-world applications [26].

In Fig. 8b, the cross-entropy loss for both training and test datasets gradually declines, approaching a low, stable value. This downward trend reflects increasing prediction accuracy and reduced uncertainty in the model's outputs. A continually

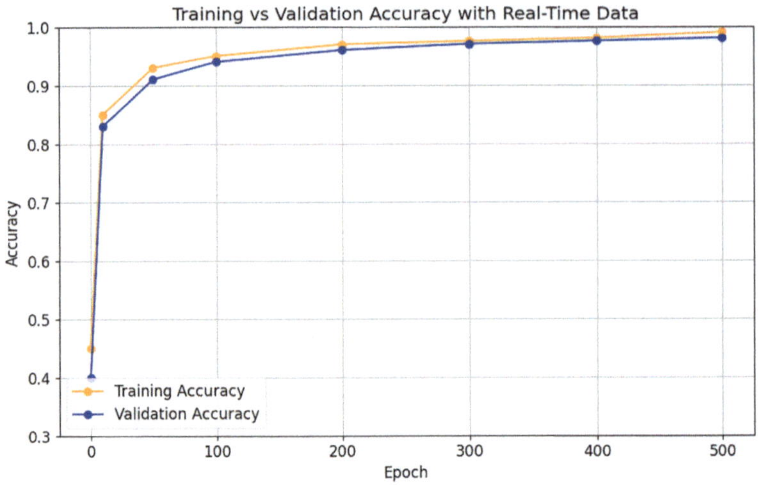

a Accuracy of the re-trained model with real-time data

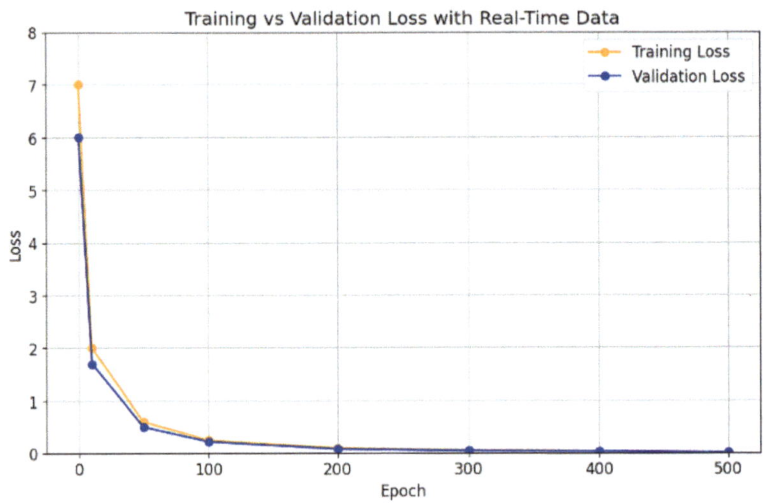

b Cross-Entropy Loss of Re-trained Model on Real-Time Data

Fig. 8 **a** Accuracy of the retrained model with real-time data. **b** Cross-entropy loss of retrained model on real-time data

decreasing loss function demonstrates that the model is learning meaningful patterns from the data. The smooth convergence of the loss curves confirms that the model is well-trained and ready for deployment in practical, real-time scenarios.

Graphical User Interface (GUI)

A graphical user interface (GUI) was built using Python's PyQt5 library [27–29]. This GUI is designed to capture real-time images from the live video stream of the product frames moving along the assembly line. It offers an easy-to-use and interactive way to view the video feed and capture frames whenever needed during the production process.

Once the inspection is initiated via the user interface, the application begins processing the input based on the programmed classification task to distinguish between frames with bullets and frames without bullets. Upon completion of the classification, the application generates the input image with red rectangular boxes that highlight the empty bullet holes. Rectangular boxes are placed surrounding the recognized empty holes based on the results of the image classification. Functions are defined to preprocess images for the model, make predictions, and display the results. Predictions are made on regions of interests (ROIs) within images, and bounding boxes are drawn around these ROIs based on the model's predictions. Regions classified as "Without bullets" are marked in red, while those classified as "With bullets" are marked in green. The final inspection stage on the assembly line, shown in Fig. 9, presents an input image of the product's frame with missing bullets. Each empty bullet hole is clearly marked with a red rectangular box, highlighting their absence as identified by the application's image classification. An alert system then notifies the assembly line operators, prompting them to take the necessary corrective actions.

IoT Cloud

The authentic data is transmitted to the cloud infrastructure for remote monitoring and subsequent actions as deemed necessary. In the proposed system, Thingspeak cloud infrastructure is harnessed to facilitate the seamless transfer of data for in-depth analysis. As depicted in Fig. 10, the product ID, along with the count of unoccupied bullet points, is visually represented as a graph. Products exhibiting an excess of 20 missing bullet points are distinctly flagged. Furthermore, the product images are conveyed to the cloud environment for comprehensive remote analysis.

Fig. 9 Final inspection in assembly line

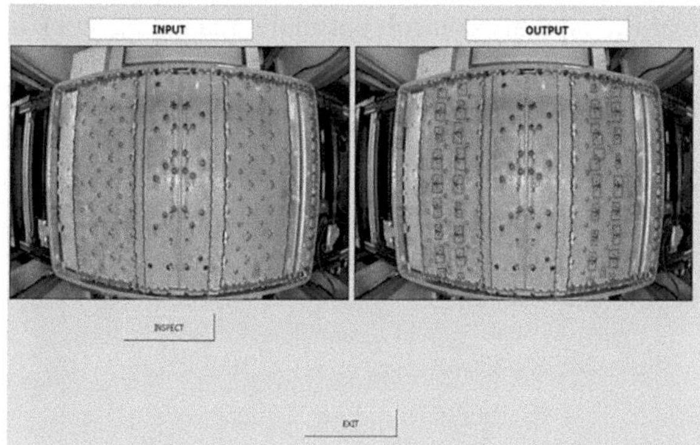

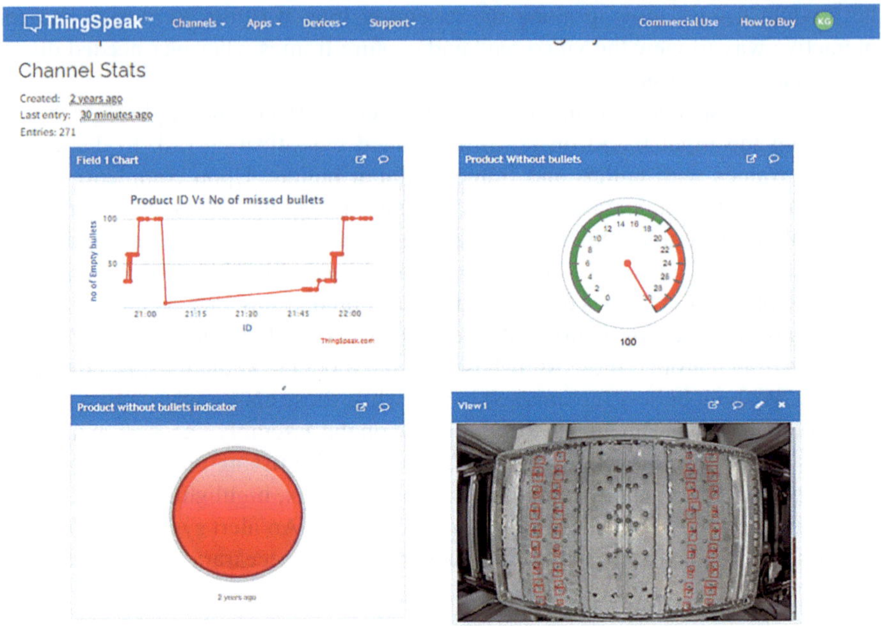

Fig. 10 IoT cloud platform for visualization of product inspection

6 Conclusion

As a computer vision-based solution and conforming to the Industry 4.0 guidelines, the identified solution has a high possibility to revolutionize logistics and supply chain management. The system can help identify the frames marked as ones with or without bullets with a high efficiency by reveling the assembly components through an image classification process. It was done with the help of a retrained MobileNet-based model that allows carrying out the high-speed and accurate recognition of the missing or existing components in real time.

This smart detection feature greatly improves efficiency on operation by enabling real-time quality control, optimally managing stock inventory, and reducing human-error. Consequently, the system leads to better decision-making and low expenses incurred in operations. Computer vision proves to be relevant in pursuing SDG 9 (Industry, Innovation, and Infrastructure) and SDG 12 (Responsible Consumption and Production). Currently, it supports automation, waste minimization, and sustainable manufacturing solutions to modern logistics. Finally, such a strategy contributes to the stability of product quality, increased productivity, and overall trust at the supply chain level.

References

1. Sandler, M., Howard, A., Zhu, M., Zhmoginov, A., & Chen, L. (2018). MobileNetV2: A lightweight CNN with inverted residuals and linear bottlenecks. arXiv preprint, https://arxiv.org/abs/1801.04381
2. Saturno, M., Pertel, V. M., Deschamps, F., & de Loures, E. F. R. (2018). Designing an automation solutions framework tailored for Industry 4.0. LogForum, 14(2). https://doi.org/10.17270/J.LOG.2018.266
3. Chandan, G., Jain, A., Jain, H., & Mohana. (2018). Deep learning and OpenCV-based real-time object detection and tracking. In International Conference on Inventive Research in Computing Applications (ICIRCA). https://doi.org/10.1109/ICIRCA.2018.8597266
4. Marengoni, M., & Stringhini, D. (2011). Using OpenCV for advanced computer vision tasks. In SIBGRAPI Conference on Graphics, Patterns, and Images Tutorials. https://doi.org/10.1109/SIBGRAPI-T.2011.11
5. Wu, J., Peng, B., Huang, Z., & Xie, J. (2013). Object detection and classification through computer vision techniques. In Lecture Notes in Computer Science, Vol. 392. Springer
6. Szegedy, C., et al. (2015). Deeper convolutional networks with improved architecture. In CVPR 2015. https://doi.org/10.1109/CVPR.2015.7298594
7. Sehgal, J., Sharma, M., Chatterjee, J., & Mehra, A. (2020). Deep learning model comparison for vehicle detection. In ICCSP 2020, Chennai. https://doi.org/10.1109/ICCSP48568.2020.9182346
8. Tripathi, S., & Kumar, R. (2019). Compact CNN models for image classification. IEEE Xplore.
9. Guennouni, S., Mansouri, A., & Ahaitouf, A. (2014). On-device multiple object detection using OpenCV. IEEE Xplore.
10. Sae-Lim, W., Wettayaprasit, W., & Aiyarak, P. (2019). MobileNet-based CNN for classifying skin lesions. In JCSSE 2019. https://doi.org/10.1109/JCSSE.20O19.8864155
11. Mohan, A., Papageorgiou, C., & Poggio, T. (2001). Component-based object detection in images. IEEE Transactions on Pattern Analysis and Machine Intelligence, 23(4), 349–361. https://doi.org/10.1109/34.917571
12. Ni, J., Khan, Z., Wang, S., Wang, K., & Haider, S. K. (2016). Circular Hough transform for overlapping circular object detection. In WCICA 2016, 2902–2906. https://doi.org/10.1109/WCICA.2016.7578268
13. Zhang, X., Zhou, X., Lin, M., & Sun, J. (2018). ShuffleNet: Highly efficient CNNs for mobile platforms. In CVPR 2018, 6848–6856
14. Sahoo, S. K., Sharma, M. M., & Choudhury, B. B. (2019). Dynamic inspection setup for bottle defects. In Computational Intelligence in Data Mining, Springer, 873–884
15. Jiang, J., Cao, P., Lu, Z., Lou, W., & Yang, Y. (2020). Surface defect recognition for smartphone glass using CNNs. Applied Sciences, 10(10), 3621
16. Bai, X., Fang, Y., Lin, W., Wang, L., & Ju, B. F. (2014). Phase spectrum-based defect detection in industry. IEEE Transactions on Industrial Informatics, 10(4), 2135–2145
17. Lin, J., Yao, Y., Ma, L., & Wang, Y. (2018). CNN-based approach for casting defect identification. International Journal of Advanced Manufacturing Technology, 97(1), 573–581
18. Srinivasan, K., Dastoor, P. H., Radhakrishnaiah, P., & Jayaraman, S. (1992). FDAS: Framework for defect analysis in woven fabrics. Journal of the Textile Institute, 83(3), 431–448
19. Vafadar, A., Guzzomi, F., Rassau, A., & Hayward, K. (2021). Metal additive manufacturing: A survey of processes and challenges. Applied Sciences, 11(3), 1213
20. Lim, S., Buswell, R. A., Le, T. T., Austin, S. A., Gibb, A. G., & Thorpe, T. (2012). Construction-scale AM process developments. Automation in Construction, 21, 262–268
21. Lee, H., Lim, C. H. J., Low, M. J., Tham, N., Murukeshan, V. M., & Kim, Y.-J. (2017). Review on laser use in AM. IJPEM-Green Technology, 4(3), 307–322
22. Gong, H., Rafi, K., Gu, H., Starr, T., & Stucker, B. (2014). Defect analysis in Ti–6Al–4V using powder-bed fusion. Additive Manufacturing, 1, 87–98

23. Taheri, H., Shoaib, M. R. B. M., Koester, L. W., Bigelow, T. A., Collins, P. C., & Bond, L. J. (2017). Powder AM defects: Types, causes, and measurement. Int. J. of Additive and Subtractive Materials Manufacturing, 1(2), 172–209
24. Parvez, M. M., Chen, Y., Karnati, S., Coward, C., Newkirk, J. W., & Liou, F. (2019). Fatigue testing for AM components. Applied Sciences, 9(16), 3226
25. Garg, M., & Dhiman, G. (2020). CNN-based defect detection on textured materials. Journal of the Institute of Electronics and Computer, 2(1), 28–38
26. Chen, J., Liu, Z., Wang, H., Nunez, A., & Han, Z. (2017). Deep learning-based fastener defect detection. IEEE Transactions on Instrumentation and Measurement, 67(2), 257–269
27. Park, J.-K., An, W.-H., & Kang, D.-J. (2019). CNN system for weld defect inspection. International Journal of Precision Engineering and Manufacturing, 20(3), 363–374
28. Li, W., Chen, C., Zhang, M., Li, H., & Du, Q. (2018). Data augmentation in hyperspectral image classification using CNNs. IEEE GRSL, 16(4), 593–597
29. Cui, W., Zhang, Y., Zhang, X., Li, L., & Liou, F. (2020). CNN-based inspection of AM parts. Applied Sciences, 10(2), 545

Health and Well-being: AI Solution for Sustainable Healthcare

Koushik Mondal, Himanshu Suyal, Gulshan Shrivastava, and Shiv Naresh Shivhare

Abstract Diabetic Retinopathy (DR) is a degenerative eye condition that can impair vision if left untreated. DR mainly affects retinal blood vessels due to elevated blood sugar levels in diabetics. Exudates (EX), microaneurysms (MA), hemorrhages (HM), and abnormal vascular growth are among the retinal abnormalities that make the clinical diagnosis of DR difficult. The specifics of automated DR detection and classification in retinal fundus images are presented and discussed here. The various components of automated diabetes retinopathy disease detection and classification are described in this chapter. The features of the illness are reviewed, along with imaging modalities and the ways in which these features have been applied in the past. We review and evaluate the state-of-the-art studies on DR detection carried out recently, including an overview of various approaches employed, associated problems, advantages, and limits. The performance measures used by the methods are also discussed in terms of significance. Deep learning techniques, particularly those exploiting Convolutional Neural Networks (CNNs) and transfer learning for feature extraction through retinal fundus scans, have surpassed traditional methods in the process of automated identification and classification of diabetic retinopathy.

Keywords Diabetic retinopathy · Computer-assisted diagnosis · Retinal fundus images · Classification · Deep learning · Medical image analysis

K. Mondal
Liverpool John Moores University, Liverpool, UK

upGrad Education Pvt. Ltd., Mumbai, India

H. Suyal · G. Shrivastava (✉)
School of Computer Science Engineering and Technology, Bennett University, Greater Noida, Uttar Pradesh, India
e-mail: gulshanstv@gmail.com; gulshan.shrivastava@bennett.edu.in

H. Suyal
e-mail: suyal.himanshu@gmail.com; himanshu.suyal@bennett.edu.in

S. N. Shivhare
School of Computer Science, University of Petroleum and Energy Studies, Dehradun, Uttarakhand, India
e-mail: shiv827@gmail.com

A. Khanna et al. (eds.), *Artificial Intelligence for Sustainable Development Goals*,
https://doi.org/10.1007/978-3-032-28217-0_6

1 Introduction

Generally, human glucose levels range from 70 to 100 mg/dL. However, exceeding this threshold (> 100 Mg/dL) may cause diabetes [6]. Diabetes leads to numerous complications, such as renal failure, dental problems, lower limb amputation, neuropathy, myocardial infarction, and diabetic retinopathy. Diabetic retinopathy (DR) is a condition related to diabetes that damages the blood vessels in the human eye retina. This damage can harm the retina's ability to function properly [46]. DR is the main reason for visual mutilation and sightlessness in adults worldwide during their working age. The disorder is marked by many retinal scratches, notably hemorrhages (HM), microaneurysms (MA), and a combination of soft and hard exudations. Vascular abnormalities and flow within the retinal tissue are indicated by exudation. DR can have major repercussions, including macular edema, retinal damage, and irreversible vision loss, if it is not identified early or is disregarded. In order to treat DR and prevent negative consequences, early detection and timely intervention are crucial. Expert ophthalmologists' physical examinations or retinal images are the primary methods used in conventional DR diagnosis. However, this depends on the availability of experts and may lead to human error. Accurate and timely diagnosis is essential for early intervention, treatment planning, along with enhanced patient outcomes.

In addition, automated systems are favorable because of their cost efficiency and scalability, making them useful for the healthcare domain at a limited cost. In recent years, multiple research projects have suggested deep learning methodologies [8, 11, 31, 51] using retinal color fundus images to detect DR. However, there is still scope for improvement, as the past literature has been extensively examined and limitations have been found. Many investigations have focused primarily on the recognition of the existence or non-existence of diabetic retinopathy, neglecting the critical assessment of its severity. They did not consider the use of small and unbalanced datasets in specific investigations. A significant difficulty lies in precisely predicting the low level of diabetic retinopathy and differentiating between moderate and low levels, mainly due to the absence of clear characteristics of the injury. Still, the implication of preprocessing methods, like contrast augmentation and noise reduction, has been abandoned in most of the research, affecting the classification accuracy. Several machine learning models suffer from the overfitting problem, so it has become essential to prevent this problem and improve the model's generalization [47].

The major objective of research is to investigate existing and state-of-the-art methodologies for the effective and efficient detection & classification of DR from images of the retinal fundus. This chapter also explores and investigates the associated challenges in automated DR detection, along with explaining and describing the characteristics of retinal fundus image modality, feature extraction, and performance assessment. The existing comprehensive and publicly available datasets have been described with their characteristics in context to validate the efficacy of the existing techniques.

Table 1 Summary of a few significant existing research works in automated diabetic retinopathy detection and performance

Author(s)	Methodology	Performance	Dataset
Nneji et al. [38]	Weighted fusion deep learning	Accuracy: 98.5%	Messidor [14]
Saini et al. [43]	Deep learning for imbalanced datasets	AUC: 0.81	IDRiD [21]
Kobat et al. [30]	Patch division-based DenseNet	Accuracy: 87.43% (five-class)	APTOS [5]
Alwakid et al. [4]	CLAHE and ESR-GAN	Accuracy: 98.7%	APTOS [5]
Pour et al. [36]	Efficient net and CLAHE	AUC: 0.945	Messidor [14]
AbdelMaksoud et al. [1]	E-DenseNet hybrid deep learning	Sensitivity: 96%	Various datasets (2006–2019)
Mansour et al. [34]	AlexNet-based deep learning	Accuracy: 97.93% (with LDA)	Kaggle fundus datasets
Gadekallu et al. [19]	PCA-firefly deep learning model	Accuracy: 97%	UCI diabetes
Menaouer et al. [35]	Hybrid deep learning approach	Accuracy: 90%	Ensemble of online datasets
Butt et al. [11]	Hybrid deep learning with transfer learning	Accuracy:89.29% (multi-class)	Various datasets
Dutta et al. [18]	CNN (VGG net)	Accuracy: 78%	Kaggle dataset
Bilal et al. [9]	U-net and deep learning	Accuracy: 97.92%	EyePACS [22]
Islam et al. [29]	CNN	Sensitivity: 98%	EyePACS [22]

2 Diabetic Retinopathy

Diabetic retinal disease (DR) is a severe eye disorder that can affect any individual diagnosed with diabetes. The condition results from injury to the retinal blood vessels. The increase in glucose in the blood may be related to diabetes and sometimes generates swollen and abnormal blood vessels. Due to this, anyone can have vision impairment or complete blindness. The whole advancement of DR can be broadly characterized into discrete phases.

1. **Mild Non-proliferative Retinopathy**: In the initial stage, small swollen areas or microaneurysms form in the eye's blood vessels. These may leak fluid into the eye, potentially causing minor changes in vision.
2. **Moderate Non-proliferative Retinopathy**: The second stage of the disease is worse than the first, as it blocks the blood vessels in the retina. This keeps oxygen and nutrients from getting to the eye, which leads to ischemic areas where blood flow is low.

3. **Severe Non-proliferative Retinopathy**: This is the third, and so far, many blood vessels have become blocked, which causes ischemia to spread throughout the eye. Neovascularisation is the process by which the body attempts to make new blood vessels.
4. **Proliferative Retinopathy**: During this later stage, blood vessels become weak and crooked in the retina, causing humor to fill the eye. The new blood vessels can become leaky, resulting in scar tissue formation, which may ultimately lead to retinal detachment and severe vision loss.

2.1 Diabetic Retinopathy Features

Timely detection and classification of DR is essential to avoid blindness. Identifying traits of diabetic retinopathy is essential to determine the stage of the disease and can be used for treatment. Consequently, an efficient DR detection model can be created by identifying and extracting these properties. Here, we present the descriptions of some significant features which are helpful for the DR identification:

1. **Microaneurysms (MAs)**: Microaneurysms (MAs) are small swellings in blood vessels that can be identified by their sac-like shape and red color. They may appear individually or in clusters, with sizes in the range [1, 3] pixels or [10, 100] μm. Microaneurysms are often the first sign of DR and occur when weak blood vessels enlarge in specific areas.
2. **Haemorrhages (HM)**: Structural problems in the walls of blood vessels, known as HE's, increase the likelihood of blood leaking out, which can lead to uneven blood flow. When blood vessels weaken, they are at risk of breaking, which can result in hematoma extravasation. Hematomas typically range in width from three to ten pixels. HEs and MAs can live together in some cases. They are called "red lesions" because they look and feel identical. The sizes of HEs can range from those of MAs to those of the Optic Disc. According to the depth of the retina and the observed contrast, they can show up as dots, blots, or flame-like patterns. A lot of the time, HEs do not have a big effect on your eyesight. Even so, seeing many blot haemorrhages can mean much ischemia, a sign of pre-proliferative retinopathy. While MAs are present, HEs are often the next sign of DR to show up.
3. **Exudates (EXs)**: Exudates are pale yellow deposits within the retina that frequently occur alongside microaneurysms (MAs) and are often easily noticeable. When blood leaks out of retinal vessels, it releases lipids and proteins from deposits outside cells. EXs come in many sizes, from small particles to bigger areas. Over time, they might change into "circinate" shapes that look like rings and have widths, usually between 1 to 6 pixels. They could have unsure edges and unclear shapes, known as "soft EXs." If they could have clear edges and bright shapes, known as "hard EXs." EXs usually show up in the back part of the fundus as bright, clear designs with much grey contrast against the dark capillaries in the eye. In addition, sometimes EXs may hurt blood vessels. Hard exudates can

make the retina thicker, which can cause macular failure, diabetic macular edema (DME), or even total blindness.

4. **Cotton Wool Spots (CWSs)**: Unlike microaneurysms and exudates, cotton wool spots are characterized by their larger size, blurred appearance, and irregular shapes with soft, indistinct edges. Thrombosis and blockages in blood vessels cause retinal infarction, which causes these spots. They show up as greyish-white spots on the nerve fiber layer because of localized ischemia that stops the flow of axoplasmic material. Getting several cotton wool spots, usually six or more in one eye, can be a sign of significant retinal ischemia, a sign of the early stage of diabetic retinopathy.
5. **The Optic Disc (OD)**: OD is the most uniform and brightest part of the normal Fundus imagery of the eye and has a round shape. It usually has a yellowish color. The center of the OD and its width tell us a lot about where the blood vessels are and where the macula is. Early detection of changes that may result in vision loss depends on identifying any issues with the structure, form, size, or area of the OD.
6. **Foveal Avascular Zone (FAZ)**: It has no retinal capillaries and is enclosed by the macula inside the fovea centralis. It has a width of roughly 0.5 mm. The FAZ's operation is unclear, and it varies for each person due to factors like age, skin tone, diet, and health conditions. It is a small spot, about 5–5.5 mm across, behind the head of the optic nerve. To find the FAZ, you must find the macula's center and the ends of the blood spots surrounding it. The FAZ is determined by the number of ships lost in a game of "6," which is incorrect. Diabetes retinopathy gets worse over time because of the loss of blood vessels in the FAZ. This makes it hard to see clearly. To prevent diabetic retinopathy, it is important to spot FAZ growth early on.
7. **Retinal Blood Vessels (RBVs)**: They include primary retinal artery or vein, along with any branches derived from them within the retina. Along with the nose and temporal branches, the artery splits into two main branches: the superior and inferior branches. These branches make a small network of capillaries that stay inside the inner nuclear layer and do not go outside it. The superior and inferior macular arteries branch into temporal arteries in the macula. In contrast, the central vein has smaller offshoots that do not extend to the fovea centralis. So, dividing the RBVs into groups and looking at how they branch out can give us important information about diseases and oddities. The unique bent vascular arcade from the OD and around the macula could indicate future problems. It can lead to DR lesions if you have diabetes for a long time. During RBV recognition, the vessels show many changes in grayscale and high local contrast, making the system more sensitive but less predictable.
8. **Neovascularization (NV)**: NV is the process in which the tiny new blood vessels form in optic disc when the retinal blood vessels lack adequate oxygen. NV is very dangerous to your eyesight because these arteries can easily break, causing bleeding, growth of scar tissue, and even the chance of retinal detachment. Number of nodal points signifies the severity of the disease. However, it is essential to differentiate between NV, intraretinal abnormalities, and big vessels inside

retina. NVs do not show leaks in fluorescence angiography (FA), but IRMAs do. When the ischemic retina is serious in NPDR or early in PDR, both NV and IRMAs occur.

2.2 *Dataset*

Fundus imaging has recently become a crucial tool for diagnosing various eye disorders, for example, glaucoma, keratoconus, DR, cataracts, and age-based macular degeneration. A fundus camera captures images that enable us to examine the inner retina, the optic nerve, blood vessels, the posterior pole, and the optic disc. For researchers of this field, accessing real public and private information has been very important. MESIDOR [15], MESIDOR-2 [16], APTOS [5], FIRE [26], STARE [27] Adam Hoover, E-OPHTHA-MA [17] and E-OPHTHAEX [18], IDRID [22], Retinopathy Online Challenge [23–25], DriDB [32], and EyePACS. These are some of the most well-known and publicly available datasets. EyePACS is one of the open-source datasets available for diagnosing eye diseases. EyePACS held a competition, Etienne Decenciere, whose main objective is to find and classify five severity stages of DR. There are various publicly available datasets, but they mostly suffer from the class imbalance issue. The class imbalance problem means the images are not evenly spread across severity classes. This difference makes it difficult to assess a machine learning model's performance. Researchers used a few tagged images from the dataset so that all classes are almost equal for the solution of this problem.

Numerous studies have utilized a combination of private & public data. Tan et al. [49] utilized a unique dataset called CLEOPATRA to detect and classify eye diseases using deep learning techniques. Gao et al. [21] noted that it is possible to obtain private data from hospitals for research purposes. Additionally, Wang et al. [50] accessed two public datasets, ROC and DIARETDB1, along with a private dataset from an eye hospital in London. Researchers use augmentation techniques or selected portions of the data to make training and testing fair when public datasets are not balanced. However, these methods can miss important information, which may lead to incorrect results.

3 Methodology

Acquisition of fundus images in real life consists of various imaging instruments and scenarios. As a consequence, this results in variable image quality because of differences in illumination, device calibration and user manipulation. Therefore the images may contain false features or noise that have hidden important retinal features in it. These differences can challenge the robustness of automated classifiers. A shallow-preprocessing is usually used to control these problems. This simplifies the subsequent analysis by suppressing spatial variation of samples and makes key structures in the image easier to perceive.

3.1 Preprocessing

Researchers commonly use various preprocessing techniques, including resizing, denoising, contrast stretching, image registration, and normalization to create a more uniform and standardized data set. These techniques can counteract noise and intra-individual variations resulting from different environmental conditions or devices used for imaging, respectively. Such inconsistencies are reduced by preprocessing to improve saliency of significant visual features, accuracy and robustness of image analysing and classifying systems.

3.1.1 Contrast Enhancement

Contrast enhancement is a key preprocessing method which highlights foreground in relation to the background. There are several ways to achieve better contrast with fundus shots. A well-known approach is to apply histogram equalization to the image data, which increases the overall contrast of an image without taking into account local variations of the histogram. For example, Zhang et al. [52] employed histogram normalization to adjust intensity. An alternative approach to this problem is either to consider small difference in the area of that pixel or is working with the Adaptive Histogram Equalization (AHE). CLAHE AHEcan be appliedin fundus images now slowly. Several studies, including those by Islam et al. [29], Nneji et al. [40], Alwakid et al. [4], and Aziz et al. [7], have utilized the CLAHE (Contrast Limited Adaptive Histogram Equalization) scheme. In these studies, the original AHE (Adaptive Histogram Equalization) method showed the greatest contrast enhancement in areas with similar intensity levels. However, CLAHE addresses this limitation effectively. Changes in the contrast between lights and darks could help reveal the retina's fine structures.

3.1.2 Color Space Transformation

This has been demonstrated as useful in enhancing image contrast, noise and normalization in such a context. Lin et al. [34], transformation to entropy images yielded more clear visual states compared to usual formats of images. Bilal et al. [9] observed that it is common to extract the green channel of fundus images because it contains better informational content and its clarity is higher. Pao et al. [43] that the addition of the green channel entropy, in addition to grayscale images, greatly improves visual quality in relation to using the standard imagery only.

3.1.3 Denoising and Normalization

Denoising and normalization are effective approaches for enhancing the quality of the images data. An example of this is the Nonlocal Means Denoising (NLM) technique employed for image denoising. But it should be mentioned that if denoising improves (noise is better removed), more refined details will be removed too, then the image clear/quality may get lower. So, there is a trade-off between how well the denoising method works and how much complex visual information is kept. We also use picture intensity normalization to make the data more uniform and avoid bias and long training times for the network. For pixel levels, this indicates that every image has a mean 0 and a std. dev. 1. It is necessary to normalize the pictures so that their intensities are all within a certain range because there could be big differences in intensities. Setting the intensities to a normal range keeps the data within reasonable limits, makes noise reduction better, and allows for successful analysis. Many research works, [1, 4, 31, 45], uses normalization.

3.1.4 Cropping and Resizing

Cropping and resizing can remove black lines and other things from the main part of the picture so that it concentrates only on the retina. Cropping in a square form removes noise and outliers while capturing the fundus's small, circular area. Cropping is a general preparation method, as explained in [1, 9, 33]. Bravo and Arbelaez [10] used two different editing methods for their studies. In one way, they cut out parts of the images so that the retina covered those parts. They utilized the second method by taking the largest image which could fit on retina. In addition, dataset may also include photos with different aspect ratios and resolutions. The photos will have a set size, so they are all identical. Different common CNN models can use the input pictures because they have been scaled to fit their needs.

3.1.5 Vessel and Optic Disc Segmentation

Recent research has focused on detecting lesions in fundus images, with various proposed solutions aimed at improving the effectiveness of the detection process. For example, Chudzik et al. [13] & Orlando et al. [41] discovered that some of the blood vessels such as branches, and the arteries can be frequently mistaken as lesions which causes false positives. To mitigate this issue, Chudzik and colleagues applied a technique known as Otsu thresholding to the green color channel to isolate the arteries from the entire image. Additionally, they applied morphology-based image-cleaning methods to suppress unwanted noise.

In a subsequent study, Imani and Pourreza [28] focused on blood vessel separation using an algorithm called MCA. This method combines various tools, including the Shearlet Transform and the Non-Subsampled Contourlet Transform. They successfully distinguished between rounded shapes, such as blood vessels, and small round spots, like exudates. Additionally, they applied image processing techniques, including "opening" and "closing," to improve contrast between blood vessels and lesions. Adem [3] has also processed approaches based on the Canny edge detector, and the circular Hough transform to locate and extract optic disc from fundus images. The optic disc can appear to be a lesion, so excluding it helped enhance the accuracy of lesion identification. Overall, authors investigated various image segmentation and processing methods to tackle the specific challenge of fundus image analysis. Their techniques clarified the images and allowed for more accurate detection of actual signs of disease.

3.1.6 Augmentation

DL has succeeded in processing data end-to-end without any preprocessing as opposed to other methods where raw data passes through separate pipelines. However, certain forms of preprocessing may improve it, particularly fundus images, as it is discussed in the section that follows. Augmentation techniques enhance the accuracy and robustness in the presence of a few and imbalanced samples. With image data, those might be rotations, translations, rescaling, shearing, flipping, and such operations. GANs are also used to create new images. Most studies reviewed apply these augmentation strategies to increase the image dataset's size and optimize the model's training process.

3.2 DR Detection and Classification Using Deep Learning

Researchers have extensively studied machine learning methods to enhance the early diagnosis of DR. There is a transition from traditional techniques, such as fluorescence angiography, to the use of digital fundus cameras. These cameras capture high-resolution images, leading to more accurate diagnoses. This section focuses on advanced systems that employ deep learning techniques. For tasks pertaining to feature extraction and classification within a supervised learning framework, the study investigates hybrid models that combine machine learning with deep learning and pure CNNs. Furthermore, it investigates Deep Neural Networks (DNNs)-based unsupervised learning techniques. [35].

In contrast, evolutionary algorithms comprise optimization methods inspired by natural processes and various metaheuristic strategies. Supervised learning is highly effective in diagnosing and classifying DR by utilizing DNN trained on labeled data to correctly assess retinal images as per the severity of the disease. These networks, trained in large-scale annotated image datasets, learn to identify patterns and attributes. These attributes can be associated with various phases of DR via hierarchical representations. After that model, tune the parameter so that the network minimizes the prediction error for accurate classification of the new imagery data. This method greatly aids in early detection, accurate classification of DR patients. Various techniques have been developed to enhance model performance, including one that combines transfer learning with GoogleNet and the ResNet-18 model. It results in a feature vector that is processed by various classifiers, for example, random forest and SVM, among others. Another technique employs two subnetworks working concurrently, combining their outputs for the final classification. U-Net models have also been applied for the segmentation of optic discs and blood vessels. More sophisticated methods utilize Inception-V3, based on transfer learning, to achieve higher accuracy.

Fundus Fluorescein Angiography, an invasive diagnostic method, may not be suitable for patients with less severe diabetic retinopathy (DR). Machine learning has been transformed by deep learning, especially CNN-based models, which provide improved automation and decision-making capabilities, particularly in image data processing. Unlike the traditional ML methods that depend on manually designed feature extraction, deep learning excels at automatically learning relevant features. It has proven particularly effective in the diagnosis of diabetic retinopathy, where CNNs are commonly exploited for classification and feature extraction. [42], applied four CNN models, i.e., Inception-v3, ResNet-50, Densenet-121, and Xception, utilizing transfer learning from the ImageNet dataset to address limited data issues. A consensus of these models provided the final predictions. [1]) brought E-DenseNet, which is a hybrid framework retaining EyeNet and DenseNet optimised for diabetic retinopathy detection (DR). Moreover, other models such as SqueezeNet and different Multilayer Deep Learning algorithms have been exploited for severity detection and lesion analysis of DR images. Furthermore, approaches as suggested by Menaouer and Xu [37] prepare high-speed and real-time classification via CNNs and deep learning methods. CNN ensemble approaches, such as that of Orlando et al. [41], to extract microaneurysms and hemorrhages in retinal images, preprocessing to enhance the image quality and to smooth out the noise. Gadekallu et al. [20] presented a five-layer deep neural network for early DR detection, that have modulated the PCA-based feature extraction and Firefly algorithm for dimension reduction, data normalization and preprocessing stages. Kobat et al. [31] proposed an automatic classification method based on the local retinal patch segmentation of input retinal images and the high classification accuracy with DenseNet201 transfer learning applied to segmented patches.

3.2.1 Deep Learning-Based Feature Extraction

1. **AlexNet-Based Feature Extraction**—Feature extraction follows by using a method such as that by Dai et al. [14], in which the model, so-called the Multi-Sieving CNN (MS-CNN), is proposed for detection of uneven microaneurysms in diabetic retinopathy. The MS-CNN model first proposes a new type of five-stage framework with an expert-guided statistical approach while interleaved learning with a deep mining deep CNN to map images to text. For feature extraction, fundus images are preprocessed and over-segmented with the Simple Linear Iterative Clustering (SLIC) algorithm [2].

2. **GoogLeNet-Based Feature Extraction**—The second approach of feature conveniently extraction is the GoogLeNet architecture. For instance, [23], proposed a deep learning method to diagnose diabetic retinopathy and macular edema from fundus images. Their proposed DL model used the architecture of Inception-V3 for detection of MAs and Hes. This model was further trained and tested on EyePACS-1 and MESSIDOR-2 datasets. Similarly, [48], presented fine-tuned deep convolutional neural network inspired by modified GoogLeNet for detecting diabetic retinopathy. Based on such model, components AI1 and AI2 were trained with modified Davis grading and manual staging, respectively. Performance of the model on manual grading was compared and found to be good.
3. **ConvNet-Based Feature Extraction**—ConvNet can also extract the characteristics and different researchers employed this structure. A DLM model is proposed by Quellec et al. [44] for the detection of multiple diseases of DR, the implemented method is a convolutional neural network for feature extraction which can identify several different stages of DR, these stages are Cotton Wool spots, Haemorrhages and micro-aneurysms, and some new phases of DR. The model used these large-scale datasets effectively from different heterogeneous sources. In addition to dataset hand segment imagery, it also enhanced the compactness by including relatively complex DR characteristics, i.e., Neovascularization, in the model.
4. **VGG-16-Based Feature Extraction**—One alternative method to enhance deep learning (DL) techniques is to extract features using the VGG-16 architecture. For example, [19], introduced a DL technique which combines a backpropagation neural network with a deep neural network leveraging VGG-16 for feature selection. Their study utilized a large-scale dataset containing 2000 images sourced from Kaggle. It used ten-fold cross-validation and a 70:30 split of the data for training. Retinal blood vessels, exudates, hemorrhages, microaneurysms, and fluid exudates are among the conditions that the trained model successfully identifies throughout the course of diabetic retinopathy (DR). They also used fuzzy c-means clustering with a suitable threshold to analyze particular DR phases.
5. **Deep CNN-Based Feature Extraction**—Researchers have investigated various methods for diagnosing diabetic retinopathy (DR) through the analysis of medical images. For example, Momeni et al. (2020) used a smart image processing method called EfficientNet B5 to help them find signs of DR in eye scans. They used images from the IDRiD and MESSIDOR databases, but first used a method called CLAHE to make the images look better. Their method worked really well, with accuracy scores of 0.932 and 0.945 on the two datasets. Chaturvedi et al. [12] subsequently devised a method capable of managing both single and multiple labels for the identification of DR utilizing the APTOS2019 dataset. Islam et al. [30] developed a model that utilized image patterns for the detection of DR. Tried different things to get better results, like changing the images with image augmentation and using methods to highlight important features while they were learning. They tested two common optimization methods, Adam and Stochastic

Gradient Descent (SGD), to see which one worked better in order to get better results.

3.3 *Unsupervised Algorithms*

Image-Analysis Image Analysis: The organizing and interpretation of image datasets of large scale is very effective when it is possible to allow patterns to emerge organically out of data without having to label the data in advance. The technique is especially beneficial in processing segmented pieces of images since it will help find meaningful interactions between various parts. The overall classification process can be simplified and made more accurate by choosing appropriate reference points at the very beginning of the process.

On the example of Mansour [36], considerable effort was made in the field of detecting diabetic eye diseases. The experiment was based on a comparative study of various methods with the aim of determining the trends of retinal images. The study design consisted of such steps as the image collection, image preprocessing, statistical model segmentation, and critical regions identification based on structural image analysis. This was followed by the extraction of important features of these scans using deep learning models during the disease detection process.

Equally, in different research, Chudzik et al. [13] developed a new strategy of identifying microaneurysms, small blood vessel leakages that are an early sign of diabetic retinopathy. Their method consisted of splitting the images on the retina into small segments and classifying each of them separately. Their approach was validated with available medical image databases in the form of a publication.

All these studies demonstrate the increased significance of advanced pattern recognition methods in medical imaging. Interestingly, methods that are unsupervised or semi-supervised and do not rely on labeled data exclusively are promising early results, frequently being more effective than conventional ones in the discovery of disease markers in medical images.

3.4 *Evolutionary Algorithms*

Mookiah et al. [39] proposed an automated algorithm that was supposed to help recognize the first symptoms of diabetic retinopathy (DR), in particular, exudates, bright spots in retinal pictures that are regarded as the most important signs of the condition. In order to make them more visible, they used different image enhancement techniques, such as contrast and adjustment of images, thus enhancing the quality of the input images. Analysis of visual patterns in the images was also used to facilitate correct diagnosis in the study.

Using this literature, Gadekallu et al. [20] presented a five-layered model that helps to detect DR at an early stage. They used a preprocessing step such as cleaning and enhancement of the images before a careful feature selection process was carried out so that only relevant attributes were passed to the model. The combination of these steps resulted in a better performance and more acceptable results.

A more modern development is a new method introduced by Kobat et al. [31] involving the division of retinal images into several (horizontal and vertical) patches of different sizes. This patch-based analysis helped the model to study localised image characteristics in a better manner, thus enabling the classification of DR stages to be achieved more precisely.

Table 1 presents the quantitative performance of recent significant research works to detect and classify automated diabetic retinopathy based on various performance metrics.

4 Performance Evaluation

The performance of a model is evaluated based on the confusion matrix, which is a standard tool in classification methods. It thoroughly analyzes the accuracy of the trained model. A two-by-two table enumerates true positives, false negatives, false positives, and true negatives. True positives are the cases in which any instance correctly predicts the DR detection.

True negatives refer to instances where the machine learning model correctly detects retinal images as not having DR. False positives occur when the model incorrectly labels images that do not have diabetic retinopathy as positive. False negatives happen when the model fails to recognize retinal images that do show diabetic retinopathy, incorrectly assigning them a negative label. Various assessment metrics can be derived from the confusion matrix to evaluate the model's performance.

1. **Accuracy**: The accuracy of the model's prediction can be calculated with Eq. 8.1 as follows. However, accuracy may be misleading in the context of unbalanced datasets.

$$\text{Accuracy} = \frac{\text{TP} + \text{TN}}{\text{TP} + \text{FP} + \text{TN} + \text{FN}} \tag{1}$$

2. **Precision**: Precision is the ratio of true positive predictions to the total number of positive predictions. It stresses how well the model can reduce false positives. It evaluates its precision in accurately recognizing positive cases and can be computed as follows:

$$\text{Precesion} = \frac{\text{TP}}{\text{TP} + \text{FP}} \tag{2}$$

3. **Recall**: The ratio of the correct positive instances to the total number of instances actually positive is the recall, also known as sensitivity. It checks the accuracy of

the models to identify all positive cases consistently and focuses on the reduction in false negatives. The recall calculation process is as follows:

$$\text{Recall} = \frac{\text{TP}}{\text{TP} + \text{FN}} \tag{3}$$

4. **F1 score**: The harmonic mean of precision and recall provides a balanced evaluation of the model's effectiveness, as illustrated below:

$$\text{F1score} = \frac{2 \times \text{Precesion} \times \text{Recall}}{\text{Precesion} + \text{Recall}} \tag{4}$$

5. **Specificity**: Specificity of the true negative instances against the actual all the negative instances. It improves recall, by noting the capability of the model to produce a negative class with low chances of false positives. Specificity is determined in the following way:

$$\text{Specificity} = \frac{\text{TN}}{\text{TN} + \text{FP}} \tag{5}$$

6. **Area Under the ROC Curve (AUC-ROC)**: This curve illustrates the relationship between the true positive rate (sensitivity) and the false positive rate (1—specificity) across various classification thresholds. A higher area under the curve (AUC-ROC) signifies better discriminative ability, while lower values indicate poorer performance across all thresholds.

5 Discussion

Active researchers have been able to find different deep learning models to enhance the diagnosis through effective classification of diabetic retinopathy. They are shifting away and leaving behind the traditional tool of examination like fluorescence angiography and moving to digital fundus camera, which takes high quality scans of the retina. There are three broad categories of deep learning strategies, namely supervised, unsupervised, and evolutionary algorithms. The focus of this chapter is on the solutions to the detection of automated diabetic retinopathy based on machine learning and deep learning techniques and a hybrid. Architectures CNNs have been effectively implemented, which include ResNet50, VGG16, DenseNet and SqueezeNet. Deep learning models are very useful for extracting features and handling complex, high-dimensional medical images when machine learning models fail. Another limitation of the past research has been that it studied binary classification without factoring in the severity of the stages of DR. The small, imbalanced datasets also plague some studies and their ability to be generalized. Proper forecasting of the initial phases of DR or delineation of the levels of severity is a challenge.

Also, preprocessing methods, noise reduction, contrast enhancement, and so on have not been heavily exploited to enhance image quality.

6 Conclusion

This paper examines the application of machine learning to the timely detection of diabetic retinopathy in a proactive manner. It brings up the issues and challenges connected with automated diagnosis and classification of this condition. The features of the DR are carefully analyzed in detail underlining the variety of its features and importance. A number of publicly available benchmark datasets are characterized. Moreover, a number of other performance evaluation metrics are also considered to evaluate the effectiveness of these techniques. The concept of digital fundus cameras (DFCs) is replacing conventional practices, and researchers are currently adopting them to acquire high-quality fundus images, thereby improving diagnostic performance and expanding their applications. It is directed toward the creation of smart systems that rely on deep learning (DL) methodology. There are several types of approaches that are discussed in this chapter, such as supervised, unsupervised and evolutionary algorithms. It assesses the availability of literature and the offered models, their contribution, success, and limitations to diagnose diabetic retinopathy with the help of various machine learning and deep learning techniques. As well, it covers the weaknesses and constraints of machine learning methods and highlights the benefits of deep learning methods in handling complex medical imaging data.

References

1. AbdelMaksoud E, Barakat S, Elmogy M (2022) A computer-aided diagnosis system for detecting various diabetic retinopathy grades based on a hybrid deep learning technique. Med Biol Eng Comput 60:2015–2038
2. Achanta R, Shaji A, Smith K, et al (2012) SLIC superpixels compared to state-of-the-art superpixel methods. IEEE Trans Pattern Anal Mach Intell 34:2274–2282
3. Adem K (2018) Exudate detection for diabetic retinopathy with circular Hough transformation and convolutional neural networks. Expert Syst Appl 114:289–295. https://doi.org/10.1016/j.eswa.2018.07.053
4. Alwakid G, Gouda W, Humayun M (2023) Deep Learning-based prediction of Diabetic Retinopathy using CLAHE and ESRGAN for Enhancement. In: Healthcare. p 863
5. (APTOS) APT-OS (2019) APTOS 2019 Blindness Detection Dataset
6. Association AD (2022) Standards of Medical Care in Diabetes—2022. Diabetes Care 45:S17–S38. doi:https://doi.org/10.2337/dc22-S002
7. Aziz T, Charoenlarpnopparut C, Mahapakulchai S (2023) Deep learning-based hemorrhage detection for diabetic retinopathy screening. Sci Rep 13:1479
8. Beevi ZS (2023) Multi-Level severity classification for diabetic retinopathy based on hybrid optimization enabled deep learning. Biomed Signal Process Control 84:104736. doi:https://doi.org/10.1016/j.bspc.2023.104736
9. Bilal A, Zhu L, Deng A, et al (2022) AI-based automatic detection and classification of diabetic retinopathy using U-Net and deep learning. Symmetry (Basel) 14:1427

10. Bravo MA, Arbeláez PA (2017) Automatic diabetic retinopathy classification. In: 13th international conference on medical information processing and analysis. pp 446–455
11. Butt MM, Iskandar DNFA, Abdelhamid SE, et al (2022) Diabetic retinopathy detection from fundus images of the eye using hybrid deep learning features. Diagnostics 12:1607
12. Chaturvedi SS, Gupta K, Ninawe V, Prasad PS (2020) Automated diabetic retinopathy grading using deep convolutional neural network. arXiv preprint arXiv:200406334
13. Chudzik P, Majumdar S, Calivá F, et al (2018) Microaneurysm detection using fully convolutional neural networks. Comput Methods Programs Biomed 158:185–192. doi:https://doi.org/10.1016/j.cmpb.2018.02.016
14. Dai L, Fang R, Li H, et al (2018) Clinical report guided retinal microaneurysm detection with multi-sieving deep learning. IEEE Trans Med Imaging 37:1149–1161. doi:https://doi.org/10.1109/TMI.2018.2794988
15. Decencière E, Zhang X, Cazuguel G, et al (2014) Messidor-2: Digital Retinal Images for Diabetic Retinopathy
16. Decencière E, Zhang X, Cazuguel G, et al (2004) Messidor: Digital Retinal Images for Diabetic Retinopathy
17. Decencière E, Zhang X, Cazuguel G, et al (2013a) E-OPHTHA-MA: A Database for Microaneurysms Detection in Retinal Images
18. Decencière E, Zhang X, Cazuguel G, et al (2013b) E-OPHTHA-EX: A Database for Exudates Detection in Retinal Images
19. Dutta S, Manideep BCS, Basha SM, et al (2018) Classification of diabetic retinopathy images by using deep learning models. International Journal of Grid and Distributed Computing 11:89–106. https://doi.org/10.14257/ijgdc.2018.11.1.09
20. Gadekallu TR, Khare N, Bhattacharya S, et al (2020) Early detection of diabetic retinopathy using PCA-firefly based deep learning model. Electronics (Switzerland) 9:. doi:https://doi.org/10.3390/electronics9020274
21. Gao Z, Jin K, Yan Y, et al (2022) End-to-end diabetic retinopathy grading based on fundus fluorescein angiography images using deep learning. Graefe's Archive for Clinical and Experimental Ophthalmology 260:1663–1673
22. Garg A, Kaur H, Thakur S, Saha A (2018) IDRID: Indian Diabetic Retinopathy Image Dataset
23. Gulshan V, Peng L, Coram M, et al (2016) Development and validation of a deep learning algorithm for detection of diabetic retinopathy in retinal fundus photographs. JAMA - Journal of the American Medical Association 316:2402–2410. doi:https://doi.org/10.1001/jama.2016.17216
24. Gulshan V. PLCMSMCWDNAVS (2016a) EyePACS: A Dataset for Diabetic Retinopathy Detection
25. Gulshan V. PLCMSMCWDNAVS, A. DMMGDCCD (2016b) Retinopathy Online Challenge
26. Hernandez-Matas C, De Ves E, Salinas V (2017) FIRE: Fundus Image Registration Dataset
27. Hoover A, Kouznetsova V, Goldbaum M (2000) STARE: Structured Analysis of the Retina Dataset
28. Imani E, Pourreza HR (2016) A novel method for retinal exudate segmentation using signal separation algorithm. Comput Methods Programs Biomed 133:195–205. doi:https://doi.org/10.1016/j.cmpb.2016.05.016
29. Islam MR, Abdulrazak LF, Nahiduzzaman M, et al (2022) Applying supervised contrastive learning for the detection of diabetic retinopathy and its severity levels from fundus images. Comput Biol Med 146:105602
30. Islam SMS, Hasan MM, Abdullah S (2018) Deep Learning based Early Detection and Grading of Diabetic Retinopathy Using Retinal Fundus Images. arXiv preprint arXiv:181210595
31. Kobat SG, Baygin N, Yusufoglu E, et al (2022) Automated diabetic retinopathy detection using horizontal and vertical patch division-based pre-trained DenseNET with digital fundus images. Diagnostics 12:1975
32. Kumar A, Poonia R, Gupta R (2021) DriDB: Diabetic Retinopathy Database
33. Li F, Wang Y, Xu T, et al (2022) Deep learning-based automated detection for diabetic retinopathy and diabetic macular oedema in retinal fundus photographs. Eye 36:1433–1441

34. Lin G-M, Chen M-J, Yeh C-H, et al (2018) Transforming retinal photographs to entropy images in deep learning to improve automated detection for diabetic retinopathy. J Ophthalmol 2018:2159702
35. Mallikarjuna B, Shrivastava G, Sharma M (2022) Blockchain technology: A DNN token-based approach in healthcare and COVID-19 to generate extracted data. Expert Syst 39:e12778
36. Mansour RF (2018) Deep-learning-based automatic computer-aided diagnosis system for diabetic retinopathy. Biomed Eng Lett 8:41–57. doi:https://doi.org/10.1007/s13534-017-0047-y
37. Menaouer B, Dermane Z, El Houda Kebir N, Matta N (2022) Diabetic Retinopathy Classification Using Hybrid Deep Learning Approach. SN Comput Sci 3:. doi:https://doi.org/10.1007/s42979-022-01240-8
38. Momeni Pour A, Seyedarabi H, Abbasi Jahromi SH, Javadzadeh A (2020) Automatic detection and monitoring of diabetic retinopathy using efficient convolutional neural networks and contrast limited adaptive histogram equalization. IEEE Access 8:136668–136673. doi:https://doi.org/10.1109/ACCESS.2020.3005044
39. Mookiah MRK, Acharya UR, Martis RJ, et al (2013) Evolutionary algorithm based classifier parameter tuning for automatic diabetic retinopathy grading: A hybrid feature extraction approach. Knowl Based Syst 39:9–22. doi:https://doi.org/10.1016/j.knosys.2012.09.008
40. Nneji GU, Cai J, Deng J, et al (2022) Identification of diabetic retinopathy using weighted fusion deep learning based on dual-channel fundus scans. Diagnostics 12:540
41. Orlando JI, Prokofyeva E, del Fresno M, Blaschko MB (2018) An ensemble deep learning based approach for red lesion detection in fundus images. Comput Methods Programs Biomed 153:115–127. doi:https://doi.org/10.1016/j.cmpb.2017.10.017
42. Oulhadj M, Riffi J, Chaimae K, et al (2022) Diabetic retinopathy prediction based on deep learning and deformable registration. Multimed Tools Appl 81:28709–28727. doi:https://doi.org/10.1007/s11042-022-12968-z
43. Pao S-I, Lin H-Z, Chien K-H, et al (2020) Detection of diabetic retinopathy using bichannel convolutional neural network. J Ophthalmol 2020:9139713
44. Quellec G, Charrière K, Boudi Y, et al (2017) Deep image mining for diabetic retinopathy screening. Med Image Anal 39:178–193. doi:https://doi.org/10.1016/j.media.2017.04.012
45. Saini M, Susan S (2022) Diabetic retinopathy screening using deep learning for multi-class imbalanced datasets. Comput Biol Med 149:105989
46. Salim SSPAPF, Ghazali KPRSAPF, Hossain MFAPF (2020) A Comprehensive Review on Automated Diabetic Retinopathy Detection. IEEE Access 8:107703–107724. doi:https://doi.org/10.1109/ACCESS.2020.3001861
47. Suyal H, Singh A (2023) Multilabel classification using crowdsourcing under budget constraints. Knowl Inf Syst. doi:https://doi.org/10.1007/s10115-023-01973-9
48. Takahashi H, Tampo H, Arai Y, et al (2017) Applying artificial intelligence to disease staging: Deep learning for improved staging of diabetic retinopathy. PLoS One 12. doi:https://doi.org/10.1371/journal.pone.0179790
49. Tan JH, Fujita H, Sivaprasad S, et al (2017) Automated segmentation of exudates, haemorrhages, microaneurysms using single convolutional neural network. Inf Sci (N Y) 420:66–76
50. Wang S, Tang HL, Hu Y, et al (2016) Localizing microaneurysms in fundus images through singular spectrum analysis. IEEE Trans Biomed Eng 64:990–1002
51. Yousef R, Gupta G, Yousef N, Khari M (2022) A holistic overview of deep learning approach in medical imaging. Multimed Syst 28:881–914
52. Zhang W, Zhong J, Yang S, et al (2019) Automated identification and grading system of diabetic retinopathy using deep neural networks. Knowl Based Syst 175:12–25

Transforming India's Path to SDGs with Artificial Intelligence: A Comprehensive Analysis

Shyam Sunder Agrawal and Amanjot Singh Syan

Abstract This chapter explores the role of Artificial Intelligence in transforming India's Path to achieving the Sustainable Development Goals (SDGs). This chapter aims to identify the role of AI-based technologies in helping industries and individuals track progress, recognize emerging challenges and patterns, and address gaps in implementation. A comprehensive analysis has been conducted in this chapter through top-cited journals from various academic databases. Artificial Intelligence has emerged as a transformative tool to address existing challenges and provide acceleration for the achievement of SDGs in India. This chapter also explores government initiatives and policies for various sectors like healthcare, education, agriculture, and environmental management that helped drive efficiency, improve decision-making, and enable more effective resource allocation. The COVID-19 pandemic has severely disrupted global efforts, with India experiencing significant setbacks, particularly in 19 of the 33 SDG indicators. As the country has worked to meet the 2030 SDG targets, integrating Artificial Intelligence (AI) in achieving all 17 Sustainable Development Goals (SDGs) offers potential benefits across multiple sectors, including society, academia, and industry. Data privacy and algorithmic bias are addressed to prevent inequalities among individuals.

Keywords Sustainable development goals · Artificial intelligence · Technology · Indian economy · and Global index

---

S. S. Agrawal (✉)
Lovely Professional University, Phagwara, Punjab, India

A. S. Syan
Canadore Stanford International College, Toronto, Ontario, Canada

A. Khanna et al. (eds.), *Artificial Intelligence for Sustainable Development Goals*,
https://doi.org/10.1007/978-3-032-28217-0_7

1 Introduction

1.1 Background of the study

Artificial Intelligence has totally transformed the path of growth of India toward sustainability, significantly contributing to achieving sustainable development goals (SDGs), and this chapter delves into a comprehensive analysis of the role of AI in transforming the path of India's development among various industries and areas [8]. Seventeen SDGs are interconnected, and all contribute to the development of nations around the globe. They fulfill all social needs and focus on education, economic growth, job creation, climate change, zero hunger, and no poverty, and provide various opportunities for society [11].

The government of India has launched various schemes and programs to achieve these goals using AI technology by 2030 [4]. AI enables growth, optimum utilization of resources, innovative solutions, and enhanced decision-making processes that help to achieve sustainability [26]. AI-based solutions like remote sensing and drones monitor environmental issues like climate change and environmental risks and help agriculture. AI-enabled computing devices help doctors reduce medical errors, and AI-based solutions provide tailored services in education, etc.

Further, AI plays an important role in the upliftment of India's GDP growth, and it can be possible by considering climate change, reducing carbon emissions, automating manual tasks, and many more [24]. AI can analyze data, identify patterns, and provide predictions for making appropriate decisions. This emerging technology significantly expanded the economy of the nation.

1.2 India's Position in the SDG Global Index

India ranks 109th out of 166 countries. As shown in Fig. 1, its SDG index score has increased year over year, and in 2023, the score was 63.99 out of 100 [32]. If we compare this growth from 2000 to 2023, the overall growth in score is 11.8, and in percent growth is 22.60%. According to the present ranking, neighboring countries have better rankings like Bhutan at 61, Sri Lanka at 93, Bangladesh at 107, and Nepal at 95 [31]. India's ranking has slipped due to wide coverage, large geographical area, diversity, and sudden climate changes [36].

While considering the status of Indian states, the Environment Report 2022 said that Jharkhand and Bihar are two states that are not prepared to achieve SDG goals [10]. Various government schemes launched, such as Pradhan Mantri Awas Yojana, Pradhan Mantri Ujjwala Yojana, Ayushman Bharat Yojna, Pradhan Mantri Mudra Yojna, Pradhan Mantri Bima Fasal Yojana, and Swacch Bharat Abhiyan, all of these schemes contribute significant progress in achieving goals, economic growth, and sustainability [15]. All Indian states have registered growth and improvement in scores. From 2018 to 2023, growth has been recorded in scores, from 57 to 71,

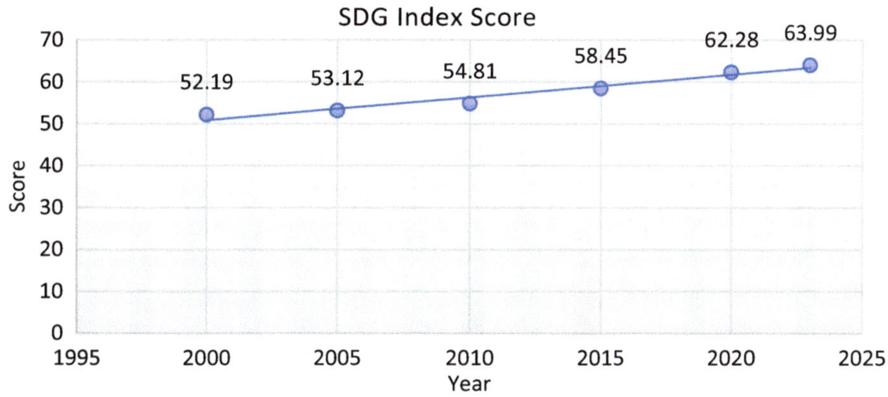

Fig. 1 SDG global index score from 2000 to 2024. *Source* Author's elaboration

and this score reflects progress in SDG 1 (No Poverty), SDG 8 (Decent work and economic growth), SDG 13 (Climate Action), and SDG 15 (Life on land). Figure 2 shows the growth of India's SDG score [13].

Further, India is committed to SDGs, NITI (National Institution for Transforming India) Aayog is focusing on the institutionalization of SDGs, it is working for sustainable development as well integral part of national and subnational thinking about development [21].

Goal-wise performance of India from 2018 to 2023–24: It provides a clear picture of growth and how technology has played an important role in this performance [20]. The Government of India focuses on various goals, especially food and nutrition security, the well-being of humans, education, smart cities, and many more [27]. Figure 3 highlights the performance of goals. In the figure, red color indicates under-performance, whereas green color indicates good performance in terms of scores [38].

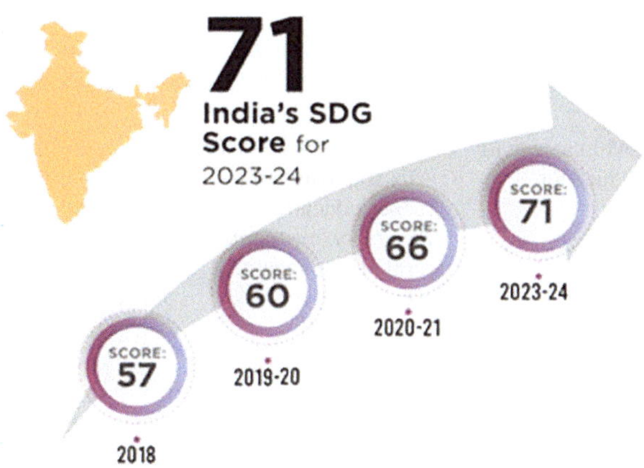

Fig. 2 India's SDG score growth from 2018 to 2023–24. *Source* https://pib.gov.in/

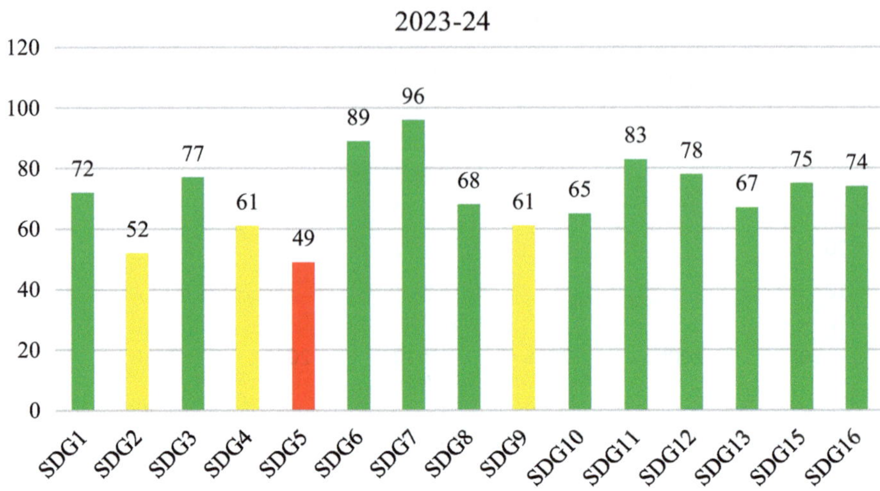

Fig. 3 Goal-wise performance of India. *Source* Author's elaboration

1.3 Research Gap

Research focus	Identified gap	Justification for bridging the gap	References
AI in economy and food security	Limited research on the role of AI in improving food security while boosting the economy through enhanced agricultural outputs	AI can optimize farming practices and food distribution, enhance productivity and addressing food security concerns. Understanding its role can improve India's agricultural and economic policies	[33, 37]
AI in access and quality of healthcare services	Inadequate focus on the application of AI in improving healthcare access in remote, rural, and economically disadvantaged regions	AI-powered healthcare solutions could reduce disparities in healthcare access, ensuring quality services for rural and underserved populations. This could enhance health equity and improve overall health outcomes	[4, 8, 14]
AI in access to education	Research lacks focus on AI's application to reduce educational inequality, especially in rural or low-income areas of India	AI-based educational tools could offer personalized learning, bridging gaps in access to quality education, and ensuring equitable opportunities for students across diverse geographical regions	[10, 12]

(continued)

(continued)

Research focus	Identified gap	Justification for bridging the gap	References
AI in mitigating discrimination	Insufficient research on AI's potential to reduce biases and discrimination in hiring, education, and public services in rural settings	AI can provide bias-free algorithms for recruitment, hiring, and policy decisions, promoting gender and social equality, especially in marginalized sectors, thereby supporting SDG 5 (Gender Equality)	[11, 16]
AI in predicting shortages and detecting inefficiencies	Limited studies on AI's capability to predict resource shortages and detect inefficiencies across various sectors in India	Predicting shortages in areas like food, water, and healthcare, and detecting inefficiencies in supply chains can enable proactive actions, enhancing India's resilience to crises like famine and economic slowdowns	[8, 17]
AI in water management	Limited research on AI's role in sustainable and efficient water management for drought-prone and water-scarce regions	AI can optimize water distribution, detect leaks, and predict water demand, contributing to efficient water use, especially in areas facing water shortages and improving water sustainability	[3, 4]
AI in environmental conservation	Lack of studies on how AI can aid in long-term environmental conservation strategies and restoration efforts in India's ecosystems	AI-based monitoring systems can aid in biodiversity protection, conservation of endangered species, and tracking deforestation and pollution, contributing to sustainable environmental practices	[11, 16, 18]
AI in economic growth and employment	A gap in research regarding AI's role in generating employment in rural areas and the impact on traditional sectors like agriculture and manufacturing	AI-driven automation and innovation in emerging fields could create job opportunities while enhancing productivity in various sectors, promoting inclusive economic growth, especially in rural India	[24, 27, 31]
AI for climate change mitigation and adaptation	Insufficient focus on how AI can contribute to real-time monitoring and mitigation of climate change impacts in vulnerable regions	AI can help improve climate prediction models, track climate-induced disasters, and provide insights for effective mitigation strategies, aiding in disaster preparedness and adaptation for vulnerable regions	[1, 8, 17]
AI in renewable energy	Lack of research on AI's role in facilitating renewable energy integration and management in off-grid, remote, and rural areas	Optimizing renewable energy systems through AI can reduce dependency on traditional energy sources, fostering sustainability, especially in remote and underserved areas	[37, 44]

(continued)

(continued)

Research focus	Identified gap	Justification for bridging the gap	References

Source Author's elaboration

2 Review of Literature

The transformation and advancements carried out by the emergence of new technology like Artificial Intelligence, machine learning, and so on. Machine learning supports attaining these goals, so much research has already been done on this [46]. AI helps in achieving SDG goals. We have discussed the role of AI in attaining these goals.

2.1 Role of Artificial Intelligence in Achieving SDGs

Artificial Intelligence has revolutionized the path of India's growth and year-on-year ranking is improving around the world. We have discussed all 17 goals defined by the United Nations. AI helps accelerate the achievement of these 17 SDGs, which are given below:

2.1.1 SDG Goal-1 (No Poverty)

UN has defined this goal to eradicate poverty around the world by the year 2030, so in this case UN has defined 7 targets and 14 indicators for this achievement because poverty arises due to lack of education, healthcare facilities, cleanliness around the city, and water sanitation [45]. According to NITI Aayog, India has recovered the poverty in the last 9 years, steep decline found in poverty headcount from 29.17 to 11.28% in the financial year 2013–14 and 2022–23, respectively. India is likely to achieve SDG target 1.2, much ahead of 2030 [36]. Significant fall in Indians living in poverty from 431 million to 129 million. Various government schemes are going on to achieve this goal like "Pradhan Mantri Garib Kalyan Anna Yojna" and "Jal Jeevan Mission," playing an important role in financial inclusion and development [48].

Artificial Intelligence (AI) has transformed the path of growth of industries and individuals through the availability of services, ease of services, proper water sanitation, cleanliness of the city, etc. AI helps to achieve this goal and it combats various factors such as lack of access to resources, income discrimination, and political instability that cause poverty [49]. AI arranges various resources like providing education, offering financial access, agricultural yield, generating employment, and improving healthcare facilities [51]. AI works on algorithms and it treats equally capable of both help and harm.

2.1.2 SDG Goal-2 (Zero Hunger)

Artificial Intelligence plays an important role in zero hunger AI helps in agricultural production and it can be possible by using AI, it helps to identify weather patterns and optimum use of resources. AI can build conversational AI to enhance better interaction with those in hunger. AI predicts optimal strategies for feeding the population using available resources, reducing waste, and improving food supply efficiency, especially in schools [42]. AI supports the agricultural output, Nutrition Early Warning System (NEWS) uses Machine Learning and Big Data to identify where risks related to food storage due to droughts, rising food prices, and crop disasters. Deep learning helps machines to identify the health of crops, sensors collect and analyze data related to crops, and IoT (Internet of Things) devices generate 4 million data points each day by 2050 [41]. AI has transformed farming activities further it will contribute to zero hunger because more production can be ensured [51]. "India is not a poor nation, yet hunger remains an issue"—India's GDP has shown growth over the last 20 years to 2.246 trillion USD but researchers say that there are still a few areas where people are facing hunger issues due to misplacement of resources [50]. India's addressing 97th rank in hunger, AI has provided a boost toward combat hunger and it also helped to improve agriculture, food, and distribution processes and predict weather conditions so that solutions can be ensured [52].

2.1.3 SDG Goal-3 (Good Health and Well-being)

To provide proper healthcare services and well-being of all age groups of citizens of the country, it focuses on child health, medicines and vaccines, and environmental diseases. The government of India has implemented several schemes for the welfare and taking care of society [46], like the Ayushman Bharat Yojana, Pradhan Mantri Jan Arogya Yojana, and Mother's Absolute Affection (MAA) [53]. National Health Mission has been working under the Ministry of Rural and Urban Health Mission to strengthen well-being and provide access to medical facilities in affordable, and accessible to all [56].

Artificial Intelligence plays an important role in the transformation of usage and accessibility of healthcare services, monitoring of patients, patient handling, and delivery of medicines. AI has reduced the time and effort to complete a particular task within the time frame [48]. AI has emerged with other technologies like nanotechnology, it has improved healthcare services and provides less expensive services related to diagnosis. Because diagnosis expenses affect directly an individual's income and savings [33]. Nanorobots also help in the treatment of patients, nanoliposomes which is used for cancer treatments and cardiovascular diseases. The advancement of technology has revolutionized the healthcare system of India [30].

According to the Global Healthcare Index 2021, India got rank 66 out of 195 nations around the world and the score is 42.8. The Indian government is planning to introduce a credit incentive program of Rs. 50,000 crore to boost the country's healthcare infrastructure [37].

2.1.4 SDG Goal-4 (Quality Education)

Education is a very important tool for the development of society, learning provides development and benefits to all human beings, and its objective is to ensure that all children get free primary and secondary education by 2030 [44]. The main aim of this is to eliminate gender discrimination in education. The Indian government has started several schemes to achieve the purpose of quality education [31].

- **SWAYAM (Study Webs of Active Learning for Young Aspiring Minds)**: The Ministry of Education launched this initiative on July 9, 2017. The main focus of this is to achieve—access, equity, and quality, through an e-learning platform. Various customized courses students, academicians, and professionals can opt for and enhance their skills, various modes available like SWAYAM portal, SWAYAM App, and SWAYAM Prabha DTH channels [20].
- **Higher Education in India—Vision 2047**: India is focusing on expanding its education infrastructure and is becoming the most favored higher education destination. India has the vision to increase the number of HEI (higher education institutions) targets and create a new student-centric ecosystem, but Artificial Intelligence can help provide skill-based programs to students [4].
- **Education for all**: The Indian government has launched several programs to achieve this aim, including Sarva Shiksha Abhiyan and Samgra Shiksha Abhiyan. For wide coverage, the government has adopted AI technology [13].
- **Project Sath-E**: Sustainable Action—Despite the successful implementation of Right to Education (RTE), NITI Aayog has initiated Project Sath-E for covering three states i.e., Madhya Pradesh, Odisha, and Jharkhand focusing on quality education [8].

Edtech companies are evolving due to customized courses and interactive lectures, gamification, and simulation-based study [1]. According to IBL News, 2021, the AI education market will touch 20 billion USD by the year 2027. AI has enhanced the user experience and helps to attain this goal [4].

2.1.5 SDG Goal-5 (Gender Equality)

This section disscuss about gender equality in all aspects like eliminating gender discrimination, initiating equal participation, and better opportunities among women and girls [3]. Various schemes launched by the government of India to eradicate this discrimination like women's helpline number (181), Beti Bachao Beti Padhao, One Stop Centre, and Kishori Shakti Yojana, and these schemes cover various aspects like providing nutrition, improvement in the sex ratio, to support women affected by violence, in private and public spaces with family member [11].

Artificial Intelligence helps identify and mitigate biases in recruitment, promotions, and other professional domains [16]. AI works on a framework that creates a healthy environment and reduces discrimination by providing equal opportunity. AI provides tailored learning platforms, bridging the gap between genders and providing education, AI promotes the accessibility of education and skill development programs for women and girls [26]. UNESCO (United Nations Educational Scientific and Cultural Organization) launched an AI-based platform named women4ethical that will improve gender equality in society [20].

2.1.6 SDG Goal-6 (Clean Water and Sanitation)

This goal ensures that providing clean water to all provides accessibility for drinking and proper sanitation; it also focuses on the supply of fresh water and adopts some practices like water conservation, etc. The Indian government has initiated various schemes related to providing clean water like the "Jal Jeevan Mission" and "Swachh Bharat Mission" to attain this goal [22]. AI helps the government with water treatment, detecting problems, and providing solutions to the government so that development can be done. AI can optimize water distribution by analyzing consumption patterns, knowing about demand and supply, and addressing various issues related to inadequate distribution [27].

2.1.7 SDG Goal -7 (Affordable and Clean Energy)

Every country is focusing on saving its renewable resources and energy because resources are limited. Artificial Intelligence helps reduce costs and improve the effectiveness of producing renewable energy, and it can be possible through robotics, predictive analysis, and ML algorithms that help to analyze the pattern and identify the results [29]. The government of India has initiated various schemes like Pradhan Mantri Urja Suraksha Evam Utthaan Mahabhiyaan, a carbon credit trading scheme, and the government has set a target of 500 GW of power from non-fossil fuels by 2030 [34]. Solar energy plants are also established to provide affordable output and contribute toward SDG and it can be possible through AI [63].

2.1.8 SDG Goal -8 (Decent work and Economic growth)

India has taken so many initiatives related to economic growth, like Startup India, Make in India, and Digital India; these programs contribute to generating employment opportunities [69]. This goal can be achieved through two parameters: urbanization and manufacturing. India is growing at a rate of 7.5% per year; rapid industrialization provides job creation, and the Maruti Suzuki plant was set up in Haryana in collaboration with Japan [66]. The government is increasing the number of MSMEs; AI supports better working conditions using robots and minimizing strain on workers, and AI supports organizations for online courses designed in a way that supports the enhancement of employee skills and their work. AI will contribute to 967 billion dollars in the Indian economy by 2035 [57].

2.1.9 SDG Goal -9 (Industry, Innovation, and Infrastructure)

India's focus on industrial growth and infrastructure development has been pivotal in achieving SDG 9. The government's flagship initiatives like "Make in India," "Digital India," and "Smart Cities" are revolutionizing industries and infrastructure. Artificial Intelligence (AI) has played a crucial role in this transformation [18]. The country has seen significant strides in manufacturing automation, with AI-enabled technologies increasing efficiency and productivity. Today, industries are being integrated into these technologies like robotics, machine learning, and the Internet of Things (IoT), especially in sectors such as automobiles, textiles, and electronic goods [23]. The AI market in India is projected to grow from $5 billion to $7.8 billion from 2020 to 2025 which will lead to innovations in production processes, its design, and supply chain management [25].

AI is also important in infrastructure development which will help to achieve this goal. Smart cities use AI and IoT to improve urban services, its traffic management, proper use of waste disposal, and energy usage [63]. The government has launched smart projects for smart cities across India, leveraging data analytics and AI has enhanced urban living standards and their conditions. Further, AI has transformed infrastructure management through predictive maintenance, analysis, and improving the reliability of infrastructure assets and their longevity [68].

Therefore, AI is working as a catalyst for India's industrial as well as infrastructural transformation, boosting the economic growth of the country and positioning the country as a leader in terms of innovation and technology [59].

2.1.10 SDG Goal-10 (Reduced Inequality)

The objective of SDG 10 is to reduce inequality within and among countries, and AI has taken a transformative action in addressing disparities related to caste, religion, and income also in India. One of the major areas where AI has made a difference in financial inclusion [65] is because a large part of the Indian population was previously excluded from the formal banking system, and AI-based solutions like mobile banking apps, e-wallets facilities, and credit score options provide more accessibility to financial services to rural and underserved areas [59]. For example, IndiaStack is a digital infrastructure framework that uses AI to provide services such as e-KYC, and AI enables access to government schemes and financial services to millions of individuals directly from their mobile phones [53].

AI also plays an important role related to healthcare and the well-being of people, enabling telemedicine and diagnostic solutions to people who live in urban and rural areas and healthcare access [46]. AI-enabled diagnostic tools like Artificial Intelligence-powered mobile health apps, self-test kits, or devices are being used to provide remote consultations, online consultations, diagnose diseases, and offer patient-wise health recommendations, thus improving healthcare access in underserved or uncovered regions [34].

Education is one of the important parameters to reduce inequality; AI has enabled personalized learning more widely accessible to students and professionals with AI-enabled platforms and helping students and professionals in remote areas to receive tailored education or personalized skill-based programs [12]. AI has broken geographical and socio-economic barriers and provided equal opportunities for all, making it easier to reduce inequality in India [22].

2.1.11 SDG Goal-11 (Sustainable Cities and Communities)

SDG 11 focuses on sustainability, and more inclusive, and resilient cities and societies, India is facing this challenge due to rapid urbanization [1]. India's urban population is increasing very fast and it is projected to reach 600 million by 2031, putting immense pressure on resources, infrastructural development, and good governance. Artificial Intelligence has changed the path of formation of smart cities that have improved urban living standards and provide proper sanitation with the help of technology [10]. The Indian government is focusing on converting smart cities so that proper sanitation can be done and it has already taken several initiatives for smart cities, for example Pune, Surat, Hyderabad, and Bengaluru have integrated AI-based solutions not only managing traffic but also waste management and enable recyclability, water conservation, and energy efficiency [18]. Indore has been recorded 8 times as the cleanest city in India because Indore Municipal Corporation has taken various initiatives related to resource management, reduced wastage, and proper recyclability [13].

2.1.12 SDG Goal-12 (Responsible Consumption and Production)

Responsible consumption and production patterns are very important for every nation because they will automatically become responsible to people for consumption and save resources for future generations. In India, AI plays a significant role in promoting responsible consumption and ensuring optimum utilization of resources across various industries [35]. The country is experiencing rapid industrial growth but it is not focusing on responsible consumption, resource depletion, and affecting the environment [21]. AI has the potential to address key issues by improving resource efficiency and reducing waste [11].

AI also helps in waste management systems to enhance recycling efforts and use automated sorting systems that will separate recyclable materials, making waste recycling more effective [38]. In the agriculture industry, today, AI-based precision farming techniques are promoting sustainable practices, reducing pesticide use, water conservation, and increasing crop production through data-driven insights. AI helps India to meet its sustainability targets and become a more responsible economy [42].

2.1.13 SDG Goal-13 (Climate Action)

Climate change affects the industry performance; due to this impact, this goal has recognized the urgent need for climate action. AI has transformed the addressing of climate challenges through better prediction, mitigation of the risk, and adaptation of various strategies [43]. AI-driven models help to analyze vast amounts of climate data to predict weather patterns and conditions, assess environmental risks, and monitor real-time changes in the environment [49].

AI is used for improving climate resilience in agriculture. It helps farmers to predict weather patterns, optimize irrigation, and monitor soil health, mitigating the risk of crop loss due to unexpected climate change [56]. Like, the Indian government launched the National Mission for Sustainable Agriculture, AI helps farmers to give solutions to climate-related challenges [61].

Moreover, AI-based technologies are crucial for mitigating carbon emissions. AI uses smart grid systems so that optimum energy can be used, integrating renewable energy sources like solar and wind into the grid more efficiently [63]. In cities, AI-enabled systems support energy management in buildings and industries, reducing overall energy consumption and carbon footprints [54].

2.1.14 SDG Goal-14 (Life Below Water)

It is focused on the conservation of natural resources like oceans, seas, and marine resources. India has a long coastline of over 7500 km: challenges related to overfishing, water pollution due to industry wastage, marine pollution due to ship garbage, and habitat degradation [60]. AI plays an important role in managing these resources and protecting whole ecosystems i.e., life below water.

AI-enabled technologies are used to monitor water's biodiversity and track fish populations. Machine learning algorithms help to analyze satellite data for identifying trends in sea surface temperatures, coral bleaching, and seawater pollution [65]. This technology helps scientists to understand the health of seawater ecosystems and predict the future trends [67]. AI is also used for various things like marine conservation efforts, by using drones and underwater robots using AI, deployed for real-time monitoring of coral reefs and underwater ecosystems [64]. AI not only supports optimizing fishing practices by providing various insights into fish migration patterns but also allows for sustainable fishing so that longevity can be created and for their survival and prevents overexploitation of marine resources. AI predicts and monitors harmful algal blooms that affect both marine life and human health; it creates longevity and survival for their lives [68].

2.1.15 SDG Goal-15 (Life on Land)

The goal aims to determine the use of terrestrial ecosystems that affect sustainability and life on land so to protect, restore, and promote sustainability, the government has taken several initiatives for this goal. India has a rich biodiversity of various species which has been given by nature as a gift, but still India has threats from deforestation, land degradation, and poaching, which can be tracked through AI [64]. AI not only supports the protection of wildlife but also monitors forest health [1] and makes sure that sustainable use can be possible of land resources [54]. Technologies like satellite imagery and remote sensing are used for monitoring forests and tracking illegal deforestation, helping to identify patterns for land cover and take timely action against this deforestation that helps the government and various stakeholders [4].

2.1.16 SDG Goal-16 (Peace, Justice, and Strong Institutions)

For any nation, it is very important that how the country is promoting peace, justice, and strong institutions [16]. AI supports to identification and detects criminal patterns, improves surveillance, is more controlling, and solves cases more efficiently [17]. AI-based systems help to ensure a reduction in corruption and create peace and a strong justice system in the country [31]. AI has made advanced justice systems that is more efficient and it has a framework that allows the government to take action against these vulnerabilities, and chatbot technology helps in this by detecting and supporting the system AI-based platforms not only cover public grievances but also provide accessibility to services [28]. Recently, the government of India has used AI for public administration, like e-governance initiatives, which have enhanced transparency as well as reduced bureaucratic inefficiencies and provided peace, and justice in the country [29].

2.1.17 SDG Goal-17 (Partnerships for the Goals)

As today's world is globalized, every nation is interconnected with its international trade and commerce, emphasizing the importance of partnerships among local and global to achieve this goal. Artificial Intelligence not only builds relationships among nations and fosters collaboration for knowledge sharing, technology transfer, and much more, but it will also address global challenges such as climate change and global warming [36]. India has developed an ecosystem that is AI-enabled and it provides active engagement in international partnerships the same [42].

India has worked with several organizations, like ASEAN, BRICS, G-20, and others, and these collaborations lead to businesses and governments that focus on global issues and challenges among nations and promote sustainability [46]. AI also offers healthcare facilities around the world, the latest therapy or technology in the field of medical or critical diseases that will help nations [50]. These partnerships also address socio-economic effects as well as regional and worldwide issues that can be resolved using artificial intelligence [1].

3 Methodology and Analysis of Status of SDG

In this chapter, we aimed to provide an in-depth analysis of India's journey toward achieving the 17 Sustainable Development Goals (SDGs) outlined by the United Nations. We have conducted a comprehensive review of available literature, research reports, and official documents to understand the strategies, policies, and programs implemented by the Government of India to address each SDG [38]. A comprehensive analysis was deployed for this study to identify the role of AI, government policies, and schemes for SDG performance. In this chapter, we have considered top-cited research articles, book chapters, and research reports through search strategy in academic databases [18]. We have considered 65 articles for this chapter to explore the role of AI in the transformation of India's path to SDGs. We have used the domain categorization method for the extraction of data and its relevancy.

In the literature review section, we have covered all 17 SDG goals in detail like various programs, schemes, or policies that have been taken by the government of India. AI plays an important role in achieving each SDG goal, Table 1 mentions in detail with roles and their status, that have transformed India's path toward growth.

4 Findings and Discussion

Artificial Intelligence plays a significant role in addressing some of the most critical challenges facing India while aligning these Sustainable Development Goals (SDGs). The integration of AI technologies is not only enhancing the efficiency and productivity of the country's industries but also helping to build a more equitable and sustainable future for the nation [19].

4.1 Effective Tool in Targeting and Delivering Welfare Resources

In the context of poverty reduction, AI is an effective tool for targeting and delivering welfare resources to the most endangered populations. AI systems not only analyze large sets of demographic and socio-economic data but also pinpoint various areas of need; therefore, they enable the government and companies to optimum distribution of resources [39]. That will ensure that financial help, welfare benefits, and social services are accessed by those who need them. Moreover, AI has revolutionized

Table 1 Role of AI in achieving SDG goals and its current status

S. No.	SDG goals	Various schemes and initiatives taken by India	Role of AI in achieving SDG goals	Current status of SDGs in India
1	SDG 1 (No poverty)	"Pradhan Mantri Garib Kalyan Anna Yojna," "Jal Jeevan Mission"	AI helps in providing education, financial access, and healthcare, and improving agricultural yield	Poverty has decreased from 29.17 to 11.28% in the last decade
2	SDG 2 (Zero hunger)	Improved agricultural schemes; "Nutrition Early Warning System"	AI supports weather predictions, optimizes food distribution, and predicts crop health	India's hunger rank is 97; AI supports better resource distribution
3	SDG 3 (Good health and well-being)	"Ayushman Bharat Yojana," "Pradhan Mantri Jan Arogya Yojana"	AI helps in medical diagnostics, patient monitoring, and healthcare optimization	India ranks 66 in global healthcare; AI is improving healthcare access and affordability
4	SDG 4 (Quality education)	SWAYAM, project Sath-E, education for all	AI enhances learning with customized courses, gamification, and skill-based programs	India is expanding its educational infrastructure with AI-driven solutions
5	SDG 5 (Gender equality)	Beti Bachao Beti Padhao, Women's helpline number	AI reduces gender biases in recruitment and education and promotes equal opportunities	AI promotes gender equality by enhancing women's access to education and skills
6	SDG 6 (Clean water and sanitation)	"Jal Jeevan Mission," "Swachh Bharat Mission"	AI optimizes water distribution and treatment, ensuring better sanitation	Efforts are ongoing to improve water accessibility and sanitation across India
7	SDG 7 (Affordable and clean energy)	Pradhan Mantri Urja Suraksha Evam Utthaan Mahabhiyaan	AI improves energy production efficiency and supports renewable energy use	India is setting targets for renewable energy; AI is helping to meet these goals
8	SDG 8 (Decent work and economic growth)	Digital India, Make in India, Startup India	AI enhances worker productivity, aids skill development, and optimizes industrial output	India is growing economically, and AI contributes significantly to employment and industrial growth
9	SDG 9 (Industry, innovation, and infrastructure)	"Make in India," "Digital India," smart cities	AI boosts manufacturing, infrastructure management, and innovation in industries	India is advancing toward a more innovative and efficient industrial framework

(continued)

Table 1 (continued)

S. No.	SDG goals	Various schemes and initiatives taken by India	Role of AI in achieving SDG goals	Current status of SDGs in India
10	SDG 10 (Reduced inequality)	India stack, mobile banking, digital financial services	AI promotes financial inclusion, improves healthcare, and offers personalized education	AI is bridging gaps in healthcare, finance, and education for underserved populations
11	SDG 11 (Sustainable cities and communities)	Smart cities initiative	AI optimizes traffic, waste management, and energy efficiency in cities	Cities like Pune and Surat are implementing AI for sustainable urban development
12	SDG 12 (Responsible consumption and production)	Initiatives in circular economy	AI supports resource efficiency, minimizes waste, and optimizes production processes	AI helps industries adopt more sustainable and responsible production models
13	SDG 13 (Climate action)	National mission for sustainable agriculture	AI aids in climate risk assessment, reduces carbon emissions, and optimizes energy use	AI-driven solutions are being used for disaster management and sustainable farming
14	SDG 14 (Life below water)	Marine conservation efforts	AI monitors marine life, prevents overfishing, and optimizes fishing practices	AI contributes to protecting India's marine resources and biodiversity
15	SDG 15 (Life on land)	Forest and wildlife protection initiatives	AI helps monitor land degradation, deforestation, and wildlife poaching	AI is used in monitoring forests and promoting sustainable agricultural practices
16	SDG 16 (Peace, justice, and strong institutions)	E-governance, AI-powered law enforcement	AI improves public services and transparency, and enhances law enforcement	AI is improving governance, accountability, and social justice in India
17	SDG 17 (Partnerships for the goals)	AI for Good Global Summit, international partnerships	AI fosters collaboration for sustainable development and addressing global challenges	India is actively engaging in global AI partnerships to promote sustainable goals

Source Author's elaboration

financial inclusion, especially in rural areas where people have less access to traditional banking infrastructure [48]. AI-based mobile apps and other digital platforms are providing marginalized groups that will give access to financial services like credit, savings, loans, and insurance to marginalized groups [44]. It has the potential to lift people out of poverty and empower individuals with the resources to enhance their financial stability and well-being. In recent years, the use of AI-enabled platforms has already contributed to a significant rise in financial inclusion, with over 150 million rural accounts opened in India [47].

4.2 Economy and Food Security

The agriculture industry has adopted AI integration to attain this objective will contribute to the Indian economy and provide food security. AI-based solutions have improved crop yield predictions, monitoring of crops, optimizing irrigation systems, and providing real-time data on weather and soil health [61]. This will help farmers to make more informed decisions, minimizing costs and improving productivity toward agriculture. AI may predict the risks of unfavorable weather conditions, like droughts, and offer early warning signals, that enable farmers to take preventive measures [59]. In addition, AI works on sustainable agricultural practices by adopting the most effective fertilization methods; therefore, it ensures optimum use of resources and minimizing environmental impact. AI-based systems enable agriculture more resilient and align with SDG 2 (Zero Hunger) which will ensure long-term food security for a growing population [69].

4.3 Access and Quality of Healthcare Services

AI has helped bridge the healthcare gap in India, particularly in rural and underserved regions facing significant challenges. AI has addressed these gaps by offering solutions that provide both access and quality healthcare services to individuals [67]. AI-based diagnostic tools provide detailed analysis of medical images generated by the latest equipment and detect various critical diseases like cancer, tuberculosis, and diabetic retinopathy at earlier stages, which significantly improves outcomes for patients. To bridge this gap between urban and rural healthcare, the government of India has taken so many initiatives such as Pradhan Mantri Jan Aarogya Yojana and it will be covered by the National Health Mission (NHM) that provide telemedicine platforms, online treatments [51], and AI also helps doctors to provide online consultations and diagnosis through video call and enhance wide coverage and accessibility to people. AI has enhanced patient experiences through geographic location and access to healthcare (Good Health and Well-Being) [53].

4.4 Access to Education

AI has changed learning environments and increased access to high-quality education in the context of the education sector. Technologies for adaptive learning have made it possible to customize educational materials for professionals and individual students, allowing for individualized learning experiences that accommodate varying learning styles and speeds [55]. AI has made it possible to create automated grading systems and virtual classes, which help address the teacher shortage, particularly in remote areas [60]. AI-driven educational platforms have already demonstrated success in improving educational outcomes in India, like NPTEL and Course Era, particularly for women and girls, who historically face barriers to education. AI's role in education is aligned with SDG 4 (Quality Education). Education is the only tool to develop society for inclusive and equitable education that will play a critical role in empowering individuals with the knowledge, and skills needed for better employment opportunities [59].

4.5 Mitigate Discrimination

AI has proven that it has eliminated discrimination toward gender inequality like hiring, recruitment, and promotion processes by ensuring fairer and more equitable decisions can be taken [65]. Corporations and institutions are utilizing AI-enabled tools to reduce discrimination based on gender, thereby promoting greater workplace diversity and equality among women and men [64]. Further, AI has enhanced women's access to essential services like healthcare, education, and financial inclusion [41], which directly empowers them. UNESCO has launched AI-based platforms like women4ethical, which is working to promote women's rights and ensure that they will get equal treatment in every field like access to opportunities in education [40].

4.6 Predict Shortages and Detect Inefficiencies

Water shortage is a major problem around the world; in India, it is also a big problem that affects sustainability and the lives of humans; therefore, AI-based technologies are used to help with water storage and its management. Water distribution systems that anticipate shortages and identify inefficiencies such as pipeline leaks have been observed and examined by AI [58]. For instance, AI-powered smart irrigation systems have improved agricultural water efficiency, decreased waste, and minimized misuse while maintaining crop health [62]. This technology supports SDG 6 (Clean Water and Sanitation) by ensuring that water resources are used effectively, sustainably, and properly. Additionally, AI is improving urban areas' personal sanitation services and garbage management, both of which are critical to achieving the goal of universal access to safe and clean water [2].

4.7 Sustainable Resources

Sustainability is very important among nations and also helps to enable sustainable resources like energy systems, which contribute to SDG 7 (Affordable and Clean Energy). Research proves that India is increasingly dependent on renewable energy sources, especially solar and wind energy, to meet its energy demands and effective supply. AI plays a significant role in finding trends as it works on real-time data [5]. AI helps to maintain this balance between energy supply and demand using the smart grid, and integrating renewables into the national grid, and reducing energy wastage and effective utilization [6]. In addition, AI has optimized energy use in buildings and industries [7], and it helps to reduce consumption and will contribute to sustainability goals. As India is more focused on renewable energy today most of the industrial production happening on this [4]. It minimizes operational costs and improves overall business efficiency will boost the country's economy in line with SDG 8 (Decent Work and Economic Growth) [3].

4.8 Rapid Industrialization

As we are living in globalization, India's rapid urbanization is happening very fast so AI plays an important role in addressing by facilitating the development of smart cities [11]. AI has been integrated into urban infrastructure to optimize traffic management, reduce pollution, and improve public services [9]. For example, AI-enabled sensors can monitor air pollution and provide data to inform policy decisions. It also works to optimum use of energy in cities, minimizing waste, ensuring efficient resource allocation to industry, and reducing irresponsible consumption. The government of India launched Swachh Bharat Abhiyan in 2015 which is aligned with SDG 11 (Sustainable Cities and Communities), aiming to create cities that are more resilient, sustainable, and livable. Indore city is the best example that has recycled its wastage and reduce pollution, clean air, and set up water treatment plants for sustainability [6].

5 Conclusion

Artificial Intelligence (AI) has great potential for accelerating India's progress toward the Sustainable Development Goals (SDGs) of the UN. It provides support to poverty reduction, starvation, poor healthcare, and environmental degradation are just a few of the systemic issues that the integration of AI technologies has begun to address. India is making great progress in several areas thanks to AI's capacity to evaluate enormous volumes of data, allocate resources optimally, and enhance the availability and effectiveness of services [22]. In line with the fundamental goals of the SDGs, this technology revolution is not only boosting economic growth but also guaranteeing more sustainable and inclusive development.

AI's benefits to poverty alleviation, for example, are already visible through more focused welfare programs and financial inclusion initiatives in rural areas. By providing access to digital financial services, AI is assisting marginalized communities in breaking free from the poverty trap. In agriculture, AI predictive analytics are transforming farming techniques, assisting farmers in adapting to changing climatic circumstances and maintaining food security [23]. AI-powered solutions are also aiding the healthcare sector by providing faster and more accurate diagnoses, making healthcare services more accessible, especially in distant areas [1].

AI has delivered competent and fair solutions in different sectors including gender equality, water management, renewable energy, and sustainable cities. AI tools are tackling gender prejudices in hiring and selection [44], thus making the workplaces more diverse [27]. Besides that, AI is directing the flow of water toward agricultural and urban usage, which in turn supports the SDG 6 goals (Clean Water and Sanitation). Tech that relies on AI is changing the world of energy by hastening the move to renewables, perfecting the distribution of energy, and minimizing waste [30]. By 2030, India's economy will be close to $1 trillion due to AI-based systems. As per the NASSCOM and McKinsey report, the agriculture sector is witnessing the adoption of AI-assisted practices by farmers, which will lead to a 20% increase in crop yield, and 30% less water consumption, thus reflecting AI's role in the advent of environmentally friendly agricultural practices. Additionally, the medical industry is likely to gain a huge benefit, as AI-based diagnostic tools are predicted to halve healthcare expenses in the next decade. In the education sector, AI potential can be seen through the impact of personalized learning systems on 72% of students benefiting from learning in rural areas, especially those with no access to qualified teachers. Moreover, AI has played a crucial role in widening the financial net, with 1.2 billion financial transactions conducted per year through AI-powered apps, resulting in rural and underserved areas getting connected to banks and financial services [42]. The global community embraced the 2030 Agenda for Sustainable Development right after the United Nations declared it in 2015.

The implications of this study are far-reaching, as AI has shown the ability to greatly accelerate progress toward the SDGs in India. However, for these improvements to be long-lasting, extensive regulatory frameworks that address ethical concerns and ensure equitable sharing of AI benefits are required [48]. Data privacy, algorithmic bias, and access to technology are all issues that must be addressed to prevent inequalities from growing [51]. Furthermore, authorities must encourage AI literacy, especially in marginalized areas, to guarantee that all residents benefit from these advancements [47].

The future of research is vast, with countless opportunities to investigate the junction of AI and sustainable development. Future research can concentrate on improving AI models to better predict and handle complex socio-economic and environmental concerns, particularly in light of India's heterogeneous population [50].

References

1. Agarwal, P., Swami, S., & Malhotra, S. K. (2024). Artificial intelligence adoption in the post COVID-19 new-normal and role of smart technologies in transforming business: a review. Journal of Science and Technology Policy Management, 15(3), 506–529.
2. Arfanuzzaman, M. (2021). Harnessing artificial intelligence and big data for SDGs and prosperous urban future in South Asia. Environmental and sustainability indicators, 11, 100127.
3. Arner, D. W., Buckley, R. P., Zetzsche, D. A., & Veidt, R. (2020). Sustainability, FinTech and financial inclusion. European Business Organization Law Review, 21, 7–35.
4. Arora, A. (2023). Agenda 2030: India's Multidimensional Approach towards Sustainability. Issue 2 Indian JL & Legal Rsch., 5, 1.
5. Awogbemi, O., Von Kallon, D. V., & Kumar, K. S. (2024). Contributions of artificial intelligence and digitization in achieving clean and affordable energy. Intelligent Systems with Applications, 200389.
6. Bag, S., Srivastava, G., Bashir, M. M. A., Kumari, S., Giannakis, M., & Chowdhury, A. H. (2022). Journey of customers in this digital era: Understanding the role of artificial intelligence technologies in user engagement and conversion. Benchmarking: An International Journal, 29(7), 2074–2098.
7. Bayly-Castaneda, K., Ramirez-Montoya, M. S., & Morita-Alexander, A. (2024). Crafting personalized learning paths with AI for lifelong learning: a systematic literature review. In Frontiers in Education (Vol. 9, p. 1424386). Frontiers Media SA.
8. Chang, Y. L., & Ke, J. (2024). Socially responsible artificial intelligence empowered people analytics: a novel framework towards sustainability. Human Resource Development Review, 23(1), 88–120.
9. Chavda, B., Pillai, R., & Srivastava, A. (2025). Sustainable Migration: Unraveling the Connections Between Urban Planning, Migration, and AI. In AI Strategies for Social Entrepreneurship and Sustainable Economic Development (pp. 47–68). IGI Global Scientific Publishing.
10. Chopra, M., Singh, S. K., Gupta, A., Aggarwal, K., Gupta, B. B., & Colace, F. (2022). Analysis & prognosis of sustainable development goals using big data-based approach during COVID-19 pandemic. Sustainable Technology and Entrepreneurship, 1(2), 100012.
11. Deshmukh, R., Jadhav, B. R., & Jain, V. (2025). Smart Digital Transformation: A Sustainable Path to Achieving the SDGs and a Green Net Zero Economy. In Digital Innovations for Renewable Energy and Conservation (pp. 241–276). IGI Global.
12. Devi, D., & Rroy, A. D. (2023). Role of artificial intelligence (AI) in sustainable education of higher education institutions in Guwahati City: Teacher's perception. International Management Review, 19, 111–116.
13. Dhamija, P., & Bag, S. (2020). Role of artificial intelligence in operations environment: a review and bibliometric analysis. The TQM Journal, 32(4), 869–896.
14. Dionisio, M., de Souza Junior, S. J., Paula, F., & Pellanda, P. C. (2024). The role of digital social innovations to address SDGs: A systematic review. Environment, Development and Sustainability, 26(3), 5709–5734.
15. ElMassah, S., & Mohieldin, M. (2020). Digital transformation and localizing the sustainable development goals (SDGs). Ecological Economics, 169, 106490.
16. Ghosh, P., & Chakravarty, T. (2024). Sustainable development goals: challenges, opportunities, and the way forward. Sustainability: Science, Policy, and Practice in India: Challenges and Opportunities, 235–241.
17. Goralski, M. A., & Tan, T. K. (2020). Artificial intelligence and sustainable development. The International Journal of Management Education, 18(1), 100330.
18. Gupta, D. A. (2019). Making India digital: Transforming towards sustainable development. Cosmos Journal of Engineering & Technology, 9(2), 1–6.
19. Gupta, S., & Degbelo, A. (2023). An empirical analysis of AI contributions to sustainable cities (SDG 11). In The ethics of artificial intelligence for the sustainable development goals (pp. 461–484). Cham: Springer International Publishing.

20. Gupta, S., Gupta, S. B., & Gupta, M. (2024). Importance of artificial intelligence in achieving sdgs in India. International Journal of Built Environment and Sustainability, 11(2), 1–26.
21. Jain, A. K., & Mishra, S. N. (2019). Role of NITI Aayog in the Implementation of the 2030 Agenda. 2030 Agenda and India: Moving from Quantity to Quality: Exploring Convergence and Transcendence, 239–254.
22. Jain, P., & Roy, A. (2024). Catalysts of Innovation: Advancing India's Future Through Investments in Basic Science. In Unleashing the Power of Basic Science in Business (pp. 91–117). IGI Global.
23. Jain, R., Prajapati, D., & Dangi, A. (2023). Transforming the financial sector: A review of recent advancements in FinTech. Available at SSRN 4380348.
24. Jaiswal, A., & Arun, C. J. (2021). Potential of Artificial Intelligence for transformation of the education system in India. International Journal of Education and Development using Information and Communication Technology, 17(1), 142–158.
25. Jatav, S. S. (2021). Does India achieve agenda 2030 targets: a multiple lens analysis. Indian Journal of Ecology, 48(4), 1217–1225.
26. Jige, S. B., & Launa, J. J. (2025). Role of Artificial Intelligence in Sustainable Development. In Achieving Sustainability with AI Technologies (pp. 311–326). IGI Global Scientific Publishing.
27. John, H. B., Agrawal, M., Nema, P., Hasan, A., Rakhra, I. K., & Singh, A. (2025). Green growth: a bibliometric analysis of digital innovation and Sustainable Development Goals (SDGs). Journal of Innovation and Entrepreneurship, 14(1), 15.
28. Kasinathan, P., Pugazhendhi, R., Elavarasan, R. M., Ramachandaramurthy, V. K., Ramanathan, V., Subramanian, S., … & Alsharif, M. H. (2022). Realization of sustainable development goals with disruptive technologies by integrating industry 5.0, society 5.0, smart cities and villages. Sustainability, 14(22), 15258.
29. Khattri, U., Sharma, P., Adarsh, A., & Kour, G. D. (2025). How Can Artificial Intelligence Enable Sustainable Development Goals?: Leveraging Technology for a Sustainable Future. In Renewable Energy and the Economic Welfare of Society (pp. 423–450). IGI Global Scientific Publishing.
30. Kulkarni, A., Joseph, S., & Patil, K. (2023). Role of artificial intelligence in sustainability reporting by leveraging ESG theory into action. In 2023 international Conference on Advancement in Computation & computer technologies (InCACCT) (pp. 795–800). IEEE.
31. Kumar, S., Talukder, M. B., & Dewangan, S. (2025). Cutting-Edge Innovations: AI and Digital Transformation in Research and Development. In Sustainability, Innovation, and Consumer Preference (pp. 29–50). IGI Global Scientific Publishing.
32. Kumar, S., Verma, A. K., & Mirza, A. (2024). Digital Transformation, Artificial Intelligence and Society. Springer.
33. Leal Filho, W., Mbah, M. F., Dinis, M. A. P., Trevisan, L. V., de Lange, D., Mishra, A., … & Aina, Y. A. (2024). The role of artificial intelligence in the implementation of the UN Sustainable Development Goal 11: Fostering sustainable cities and communities. Cities, 150, 105021.
34. Mahajan, A., Vaidya, T., Gupta, A., Rane, S., & Gupta, S. (2019). Artificial intelligence in healthcare in developing nations: The beginning of a transformative journey. Cancer Research, Statistics, and Treatment, 2(2), 182–189.
35. Mahida, D. P., Sendhil, R., & Ramasundaram, P. (2021). Millennium to the sustainable development goals: Changes and pathways for India. Business Strategy & Development, 4(2), 136–147.
36. Mhlanga, D. (2021). Artificial intelligence in the industry 4.0, and its impact on poverty, innovation, infrastructure development, and the sustainable development goals: Lessons from emerging economies?. Sustainability, 13(11), 5788.
37. Mishra, H., & Maheshwari, P. (2020). Achieving sustainable development goals through Fourth Industrial Revolution: an Indian perspective. Indian Journal of Commerce and Management Studies, 11(2), 63–75.
38. Misra, S. K., Sharma, S. K., Gupta, S., & Das, S. (2023). A framework to overcome challenges to the adoption of artificial intelligence in Indian Government Organizations. Technological Forecasting and Social Change, 194, 122721.

39. Mor, S., Madan, S., & Prasad, K. D. (2021). Artificial intelligence and carbon footprints: Roadmap for Indian agriculture. Strategic Change, 30(3), 269–280.
40. Okunlaya, R. O., Syed Abdullah, N., & Alias, R. A. (2022). Artificial intelligence (AI) library services innovative conceptual framework for the digital transformation of university education. Library Hi Tech, 40(6), 1869–1892.
41. Olawade, D. B., Wada, O. Z., David-Olawade, A. C., Fapohunda, O., Ige, A. O., & Ling, J. (2024). Artificial intelligence potential for net zero sustainability: Current evidence and prospects. Next sustainability, 4, 100041.
42. Oliver, N. (2022). Artificial intelligence for social good: the way forward. Science, Research and Innovation Performance of the EU 2022 Report, 604-707.
43. Palomares, I., Martínez-Cámara, E., Montes, R., García-Moral, P., Chiachio, M., Chiachio, J., ... & Herrera, F. (2021). A panoramic view and swot analysis of artificial intelligence for achieving the sustainable development goals by 2030: progress and prospects. Applied Intelligence, 51, 6497–6527.
44. Pandey, A., & Shukla, S. P. (2022). Functioning of NITI Aayog of India: A Study.
45. Pandey, M., Gusain, D., & Sharma, S. (2022). Role of renewable energy in attaining sustainable development. In Artificial Intelligence for Renewable Energy Systems (pp. 69–79). Woodhead Publishing.
46. Parija, S. C., & Shrivastava, S. (2021). realizing Sustainable development Goals through research and innovation: Strategies and Approaches. UNIVERSITY NEWS, 59, 93.
47. Pigola, A., Da Costa, P. R., Carvalho, L. C., Silva, L. F. D., Kniess, C. T., & Maccari, E. A. (2021). Artificial intelligence-driven digital technologies to the implementation of the sustainable development goals: A perspective from Brazil and Portugal. Sustainability, 13(24), 13669.
48. Prabhakar, A. C. (2025). AI Strategies for Sustainable Development Goals: Collective Action for Poverty Alleviation. In AI Strategies for Social Entrepreneurship and Sustainable Economic Development (pp. 393–442). IGI Global Scientific Publishing.
49. Pullamaraju, P. (2024). Charting India's Path in the Global Technology Landscape A Comparative Policy Study (Doctoral dissertation, School of Public Policy in Partial Fulfilment of the Requirement for the Degree of Master of Public Policy (MPP) 2021-23 Pranitha Pullamaraju HP22PPOL0100026 Under the Supervision of Dr Kanica Rakhra Assistant Professor Kautilya School of Public Policy, Gandhi Institute of Technology and Management).
50. Rafi, S., Balasani, R., Qadir, F., Tahir, M., Ahmed, F., & Ahmad, W. S. (2022). Role of artificial intelligence in agriculture sustainability, an example from India. In Towards sustainable natural resources: Monitoring and managing ecosystem biodiversity (pp. 275–293). Cham: Springer International Publishing.
51. Rahman, H., D'Cruze, R. S., Ahmed, M. U., Sohlberg, R., Sakao, T., & Funk, P. (2022). Artificial intelligence-based life cycle engineering in industrial production: a systematic literature review. IEEE Access, 10, 133001–133015.
52. Rahman, M. S., Kumar, A., & Gupta, R. (2024). India's transition to a sustainable energy future: Challenges and opportunities. Water and Energy International, 67(5), 50–54.
53. Raman, R., Lathabai, H. H., Mandal, S., Das, P., Kaur, T., & Nedungadi, P. (2024). ChatGPT: Literate or intelligent about UN sustainable development goals?. Plos one, 19(4), e0297521.
54. Rane, N. (2023). Roles and challenges of ChatGPT and similar generative artificial intelligence for achieving the sustainable development goals (SDGs). Available at SSRN 4603244.
55. Rautela, R., Kumar, S., Pandey, S., Prakash, N., Malik, P. K., & Kumar, A. (2023). Significance of Emerging Technological Advancements in Transition of Green Economy. In 2023 IEEE Devices for Integrated Circuit (DevIC) (pp. 221–224). IEEE.
56. Saba, C. S., & Ngepah, N. (2024). The impact of artificial intelligence (AI) on employment and economic growth in BRICS: Does the moderating role of governance Matter?. Research in Globalization, 8, 100213.
57. Saha, S., & Shaw, P. (2019). Revisiting industrialisation and innovation in India: Roadmap for SDG 9. 2030 Agenda and India: Moving from Quantity to Quality: Exploring Convergence and Transcendence, 41–64.

58. Sahil, K., Mehta, P., Bhardwaj, S. K., & Dhaliwal, L. K. (2023). Development of mitigation strategies for the climate change using artificial intelligence to attain sustainability. In Visualization techniques for climate change with machine learning and artificial intelligence (pp. 421–448). Elsevier.
59. Saini, N., & Kharb, R. (2025). Empowering sustainable development through digital transformation: insights from digital India. Journal of Asia Business Studies.
60. Samanvitha, C., Dharmendra, H., Nethravathi, N., & Velmurugan, P. R. (2025). Artificial Intelligence for Sustainable Futures: Navigating Social and Environmental Responsibility. In Advancements in Intelligent Process Automation (pp. 129–154). IGI Global.
61. Sanders, C. E., Mayfield-Smith, K. A., & Lamm, A. J. (2021). Exploring twitter discourse around the use of artificial intelligence to advance agricultural sustainability. Sustainability, 13(21), 12033.
62. Shaikh, T. A., Rasool, T., & Lone, F. R. (2022). Towards leveraging the role of machine learning and artificial intelligence in precision agriculture and smart farming. Computers and Electronics in Agriculture, 198, 107119.
63. Sharma, R., & Gupta, H. (2024). Leveraging cognitive digital twins in industry 5.0 for achieving sustainable development goal 9: An exploration of inclusive and sustainable industrialization strategies. Journal of Cleaner Production, 448, 141364.
64. Singh, A., Kanaujia, A., Singh, V. K., & Vinuesa, R. (2024). Artificial intelligence for Sustainable Development Goals: Bibliometric patterns and concept evolution trajectories. Sustainable Development, 32(1), 724–754.
65. Singh, U. K. (2024). Implementation of Sustainable Development Goals in India: Progress and Challenges. Regional Dimensions of Human Development in India and South Africa: Through Sustainable Development Goals, 39–91.
66. Srivastava, A., & Maity, R. (2023). Assessing the potential of AI–ML in urban climate change adaptation and sustainable development. Sustainability, 15(23), 16461.
67. Strielkowski, W., Grebennikova, V., Lisovskiy, A., Rakhimova, G., & Vasileva, T. (2024). AI-driven adaptive learning for sustainable educational transformation. Sustainable Development.
68. Wahl, B., Cossy-Gantner, A., Germann, S., & Schwalbe, N. R. (2018). Artificial intelligence (AI) and global health: how can AI contribute to health in resource-poor settings?. BMJ global health, 3(4), e000798. www.https://pib.gov.in/
69. Yigitcanlar, T., Mehmood, R., & Corchado, J. M. (2021). Green artificial intelligence: Towards an efficient, sustainable and equitable technology for smart cities and futures. Sustainability, 13(16), 8952.

An AI-Driven Non-invasive Model for Diabetes Risk Prediction Using Tongue Images: A Comparative Study

G. Anitha, Mohamed Ismail, and K. S. Margaret

Abstract Diabetes mellitus (DM) and its associated symptoms, especially diabetic retinopathy (DR), remain a major health issue today. Historically, the detection of these conditions has been undertaken by invasive blood sampling and special scanning, both of which can be painful and sometimes intimidating to patients. In order to avoid such problems, researchers have started considering other methods which is easier and safer to detect. One promising solution involves the utilization of tongue photographs to assess the presence of DM and non-proliferative diabetic retinopathy (NPDR). In this study, tongue images were taken from 90 subjects (healthy volunteers, diabetics and people with NPDR). A sophisticated digital imaging system was used to obtain the camera images, ensuring consistent lighting to ensure the results were reliable. Images obtained were pre-processed such as noise cancellation, color balance, and image segmentation to extract specific features. The high-resolution images were then analyzed using machine learning (ML) techniques including discriminant analysis, K-nearest neighbors (KNN), support vector machine (SVM), naive Bayes (NB) and neural networks. Deep learning (DL) techniques, in particular convolutional neural networks (CNNs), have begun to surpass traditional ML based algorithms for medical image analysis. A CNN was also trained in this study to distinguish diabetes using tongue images. The evaluation of the performance of CNNs compared to traditional ML models was subsequently carried out to find an optimal classifier.

G. Anitha (✉)
Electronics and Instrumentation Engineering, B S Abdur Rahman Crescent Institute of Science and Technology, Chennai, India
e-mail: anidhag@crescent.education

M. Ismail
Electronics and Communication Engineering, B S Abdur Rahman Crescent Institute of Science and Technology, Chennai, India
e-mail: mmismail@crescent.education

K. S. Margaret
Electrical and Electronics Engineering, S. A. Engineering College, Chennai, India
e-mail: margaret@saec.ac.in

A. Khanna et al. (eds.), *Artificial Intelligence for Sustainable Development Goals*,
https://doi.org/10.1007/978-3-032-28217-0_8

Keywords Diabetes mellitus (DM) · Non-proliferative diabetic retinopathy (NPDR) · Tongue geometry features · Machine learning · Deep learning · Convolutional neural network

1 Introduction

In general, diabetes develops when blood sugar remains high for a long time. It is typically classified into three types: type 1, type 2, and gestational forms, with the former two types being the most common. The traditional method for identifying diabetes is the Fasting Plasma Glucose (FPG) test, in which patients are asked to fast and then have their blood drawn—a procedure that is both invasive and uncomfortable.

DR and NPDR are detected through imaging techniques like fluorescein-based retinal imaging and fundus imaging, both considered to be invasive techniques. Fundus imaging involves exposing direct illumination of the retina using high intensity light sources, while fluorescein angiography requires the injection of fluorescein dye into a vein. To avoid these invasive procedures, alternative non-invasive methods using features from tongue images has been proposed for detecting DM and NPDR.

DM can be diagnosed using advanced methods suggested by various researchers. Shu Ting et al. [1] developed a technique utilizing facial texture blocks characteristics through the Gabor filter for DM identification. N.V. Cibin et al. [2] obtained 32 attributes, comprising four energy features via wavelet decomposition, to distinguish between non-diabetic and diabetic subjects, as well as NPDR and non NPDR.

Sujatha and Divya [3] applied a median filter to pre-process the tongue images and employed a Probabilistic SVM classifier that attains higher accuracy compared to conventional SVM classifier. A non-invasive approach is presented [4, 5] for detecting DM and NPDR, extracting thirteen attributes related to measurement, distance, area, and their ratios from tongue images. Their approach integrated chromatic, textural, and geometric analysis of the tongue for the assessment of patient's health and categorize them into three distinct groups.

Thirunavukkarasu et al. [6] utilized a convolutional neural network (CNN) for classifying DM, achieving a superior accuracy rate of 94.28% compared to other classifiers. Their approaches to analyzing and classifying diabetes were based on thermal variations in the human tongue. Vocaturo et al. [7] explored prototype mobile technologies for diagnosing the tongue powered by artificial intelligence. Festin et al. [8] conducted a study to develop a portable equipment for detecting DM using a CNN, deployed on a Raspberry Pi 3 Model B and programmed primarily in Python.

Jiang et al. [9] presented their study insights on different CNN models applied in the process of selecting tongue image quality. Their results suggest that employing DL techniques, particularly deep CNNs, is effective for assessing low-quality tongue images. A DL-driven disease diagnostic model, named IDLDD-BTI is implemented [10], which utilizes a Fuzzy-based Adaptive Filtering method for removing artifacts from biomedical tongue images.

Li et al. [11] used two separate datasets of tongue images to predict diabetes: one dataset is trained in ML classifiers, the other dataset is trained using DL classifiers. The results show that the tongue attributes greatly facilitate the classification accuracy of ML models, whereas high recall and minimal false negatives are achieved by DL models. They showed significant correlation of glucose metabolism and tongue image features, and pointed out tongue data as a potential biomarker to diagnose prediabetic and diabetic conditions in early stages. Using tongue image datasets [12], the authors used ResNet-50 in order to extract subtle features, and the Tongue Diagnosis Analysis System (TDAS) for detecting color and texture traits. Then, two sets of features were integrated with Genetic Algorithm with Extreme Gradient Boosting. The developed model demonstrates strong non-invasive prediction for diabetes risk, discriminating at both prediabetic and diabetic stages. Bharathi et al. [13] applied preprocessing and feature extraction with different standard classification algorithms and classification of diabetic tongue images. The proposed algorithm based on CNN for dimension reduction has the advantage of higher accuracy than several other classification algorithms with this mechanism. Jun Li [14] studied the incorporation of unsupervised, self-supervised, and supervised learning and established a holistic model to classify diabetic tongues. The algorithm operates without human intervention, taking decision solely based on tongue images, with state-of-the-art performance. Hence, their study seeks to enhance the benefits of conventional Chinese medicine (TCM) for personalized diabetes care, with implications for the evolution of TCM-associated diagnosis toward standardization. Bharani et al. [15] used CNN for DM detection with 98% accuracy. Then, they use a modified five-layer algorithm based on the VGG16 model to classify healthy tongue from diabetic tongue. Zhang et al. [16] proposed an innovative approach by profiling a variety of color features in the tongue. Hassoon's et al. [17], which trained more than 5000 tongue photos classified by color, evaluated various ML models including Decision Trees, SVM, Naïve Bayes, KNN, Random Forest, and XGBoost. Amazingly, XGBoost was 99% accurate, which suggests that tongue-based diagnostics are remarkably promising even under conditions with poor lighting. Overall, based on these results, it was concluded that ML and DL could provide a reliable, noninvasive alternative to the conventional method for diabetes detection by performing tongue photographs by analytically visualizing tongue images, non-invasively. By comparing classical ML methods with modern CNNs, this work will attempt to investigate which is more reliable and accurate.

2 Data Collection and Processing

Participants for this study were divided into three groups which are healthy, diabetes and NPDR participants. Tongue images were taken of participants from each group by high-quality cameras with constant illumination. There are some aspects enhancing image quality that are incorporated into the pre-processing workflow:

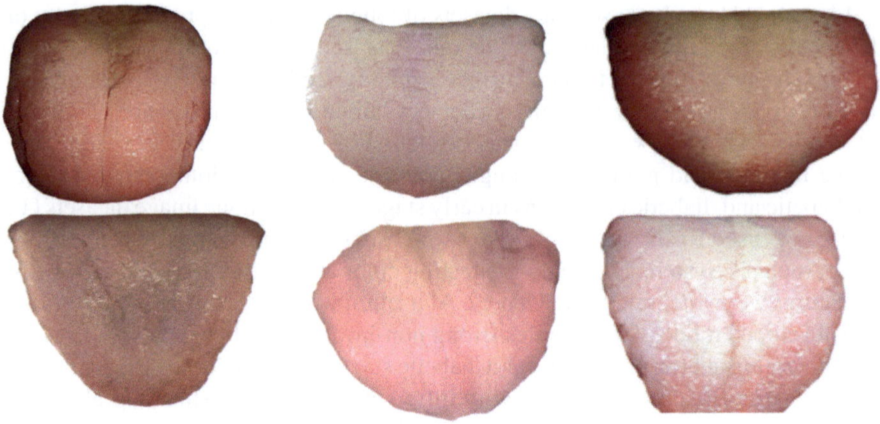

Fig. 1 Pre-processed tongue images

1. Noise Reduction: We used selective filtering to remove noise from the image and maintain important diagnostic characteristics.
2. Illumination Standardization: For the dataset, the images were given a thorough color correction to maintain uniform lighting levels across the images.
3. Automated Segmentation: We built an algorithm to separate the tongue zone from the adjacent facial section.

After noise was removed, the tongue region was segmented with masks created according to MATLAB's Color Thresholder application, as seen in Fig. 1.

3 Tongue Geometry Features Extraction

The geometric analysis of tongue images involves quantifying 13 distinct features. The parameters include basic dimensional measurements and derived ratios that characterize the tongue's shape and proportions. These features are systematically evaluated through MATLAB's measurement tools for each subject. The exact formulas used for each measurement are presented in Table 1, which details the mathematical framework underlying those formulas.

where

- P_{max} represents the rightmost edge point.
- P_{min} indicates the leftmost edge point.
- Q_{max} denotes the lowest edge point.
- Q_{min} marks the highest edge point.

The geometric features are illustrated through the visual representations provided in Figs. 2, 3, 4, 5 and 6.

Table 1 Formulae to calculate the geometrical features

SI. No	Parameter	Formula to calculate the parameter
1	Width (w)	$P_{\max} - P_{\min}$
2	Length (l)	$Q_{\max} - Q_{\min}$
3	Length-width ratio (lw)	$lw = \frac{l}{w}$
4	Smaller half distance (k)	$k = \frac{\min(l,w)}{2}$
5	Centre distance (cd)	$cd = \frac{\max(y_{\max})+\max(y_{\min})-y_{cp}}{2}$ where $y_{cp} = \frac{y_{\max}+y_{\min}}{2}$
6	Centre distance ratio (cdr)	$cdr = \frac{cd}{l}$
7	Area (a)	$a =$ number of tongue foreground pixels
8	Circle area (ca)	$ca = \pi r^2$
9	Circle area ratio (car)	$car = \frac{ca}{a}$
10	Square area (sa)	$sa = 4r^2$
11	Square area ratio (car)	$sar = \frac{sa}{a}$
12	Triangle area (ta)	$ta =$ area of a triangle defined within the tongue foreground
13	Triangle area ratio (tar)	$tar = \frac{ta}{a}$

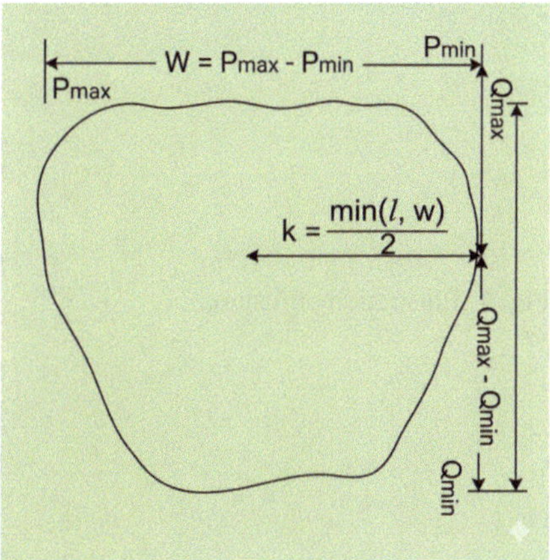

Fig. 2 Illustration of features *w*, *l* & k

4 Results

Feature extraction is the process of getting valuable image-derived information presented in numerical form.ML classifiers are effectively employed for categorizing biomedical images. The parameters derived from the tongue images, which

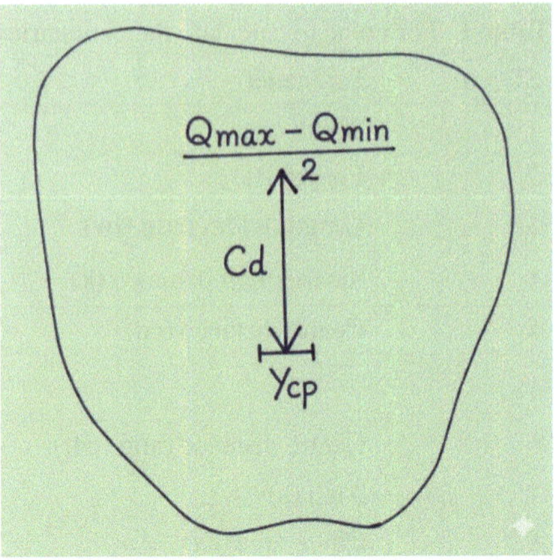

Fig. 3 Illustration of feature cd

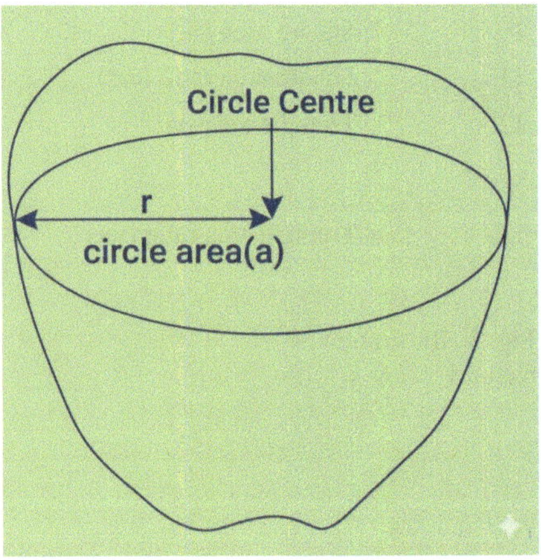

Fig. 4 Illustrationof feature *car*

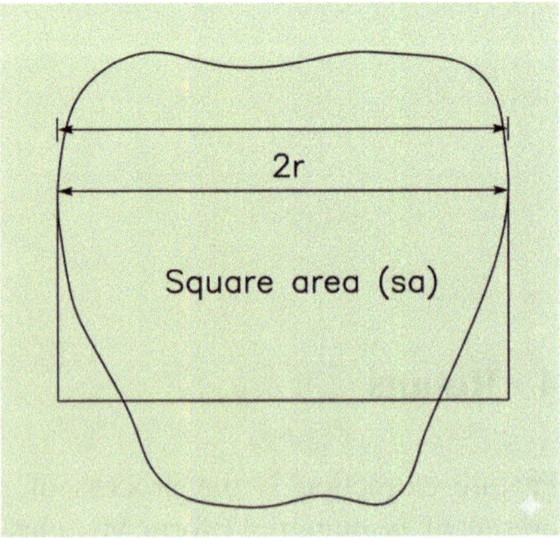

Fig. 5 Illustration of feature *sa*

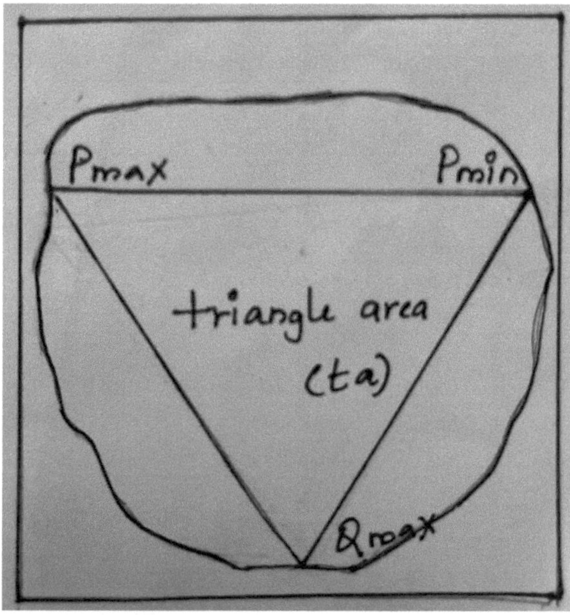

Fig. 6 Illustration of feature *ta*

are essential for making predictions or diagnoses, are input into the ML classifier. Once features are extracted from tongue images, they reveal information about the geometrical variations in the tongue images. Consequently, the data obtained from this quantification process is utilized for screening, analysis, and diagnostic purposes. Thirteen features are measured for 3 individuals and the measured features represented in pixels and summarized in Table 2.

The computed features from the tongue images are input into various ML classifiers, including SVM, KNN, NN, Ensemble and Naive Bayes. The efficiency of these different ML classifiers has been estimated using a number of tongue images The performance of the classifiers has been evaluated using metrics including accuracy, specificity, sensitivity, and precision.

Manually extracting features from a large dataset is a labor-intensive and time-consuming process. As the volume of input data increases, traditional ML models perform poorly in terms of efficiency. At this stage, the DL classifier becomes particularly relevant. As previously stated, automatic feature extraction through DL significantly reduces the time spent on manual extraction. This study employs a multilayered Deep Neural Network (DNN), capable of end-to-end layer-wise training. This approach enables the DNN to learn features that best fit the training dataset. The pre-processed tongue images are input into the DNN, and the performance is presented in Table 3.

Each ML method has its own strengths and limitations. The evaluation of the best ML classifier for the dataset is to evaluate the performance of several different types. Individual classifiers perform differently with regards to data for the tasks performed by each, and in terms of making proper decisions for a specific dataset, the choice of

Table 2 Estimated parameters for three persons

Sl.No	Features	Subject 1 (healthy)	Subject 2 (DM)	Subject 4 (NPDR)
1	Width	148	255	218
2	Length	223	201	165
3	Lw ratio	1.50676	0.78824	0.9515
4	z	74	127.5	109
5	Cd	−111.5	−100.5	−82.5
6	Cdr	−0.92	−0.43	−0.48
7	Area	33,004	51,255	35,970
8	Ca	17,195	51,045	37,306
9	Car	0.5210	0.9959	1.0372
10	Sa	21,904	65,025	47,524
11	Sar	0.6637	1.2687	1.3212
12	Ta	16,428	48,769	35,643
13	Tar	0.4978	0.9515	0.9909

Table 3 Performance analysis of ML versus DL models

Classifier	Accuracy (%)	Sensitivity (%)	Specificity (%)	Precision (%)
Discriminant	95.56	95.56	97.85	95.56
Naïve Bayes	93.33	93.33	96.77	93.33
SVM	96.67	96.67	98.38	96.67
kNN	95.56	95.56	97.85	95.56
Ensemble	97.78	97.78	98.91	97.78
NN	96.67	96.67	98.38	96.67
DNN	98.89	98.89	99.45	98.89

the appropriate classifier is very important. The DNN model outperforms competing classifiers in specificity, sensitivity, accuracy, and precision metrics. The comparative illustration of various ML models is shown in Fig. 7.

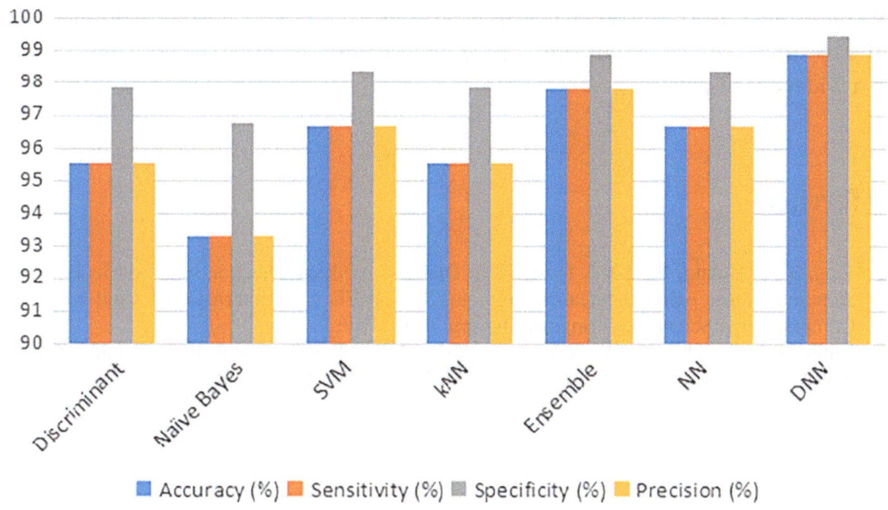

Fig. 7 Detection of DM-performance indices of classifiers

5 Conclusion

This report presents a competitive comparison between different ML and DL classifiers in order to be able to identify diabetes accurately. The tongue images of 30 non-diabetic, 30 diabetic and 30 NPDR subjects were acquired using digital imaging. After segmentation, several geometric features were extracted from all the tongue images. These extracted features are used in classification through ML and DL classifiers. Results were then analyzed to determine the best classifier. The overall effectiveness of the DNN model was evaluated, scoring a 98.8% accuracy using performance metrics comparing against other classifiers. This study leads to creation of diagnostic instruments for ease-of-use and high reliability that remains consistent with the practice standard. The non-invasive diabetes risk prediction model introduced in this study, represented an effective use of ML methods in the medical field. The results support that tongue-related data is a promising biomarker to use for early detection of diabetes. Data volume limitation continues to exist despite rigorous data collection and pre-processing efforts.

A number of stabilization strategies were adopted to maintain the consistency of features including unified lighting systems and high-resolution imaging sensors for tongue diagnosis instrument as well as standards in the software and standardizing the procedural techniques in the processing for the image analysis. Nevertheless, the consistency of tongue features may be influenced by image storage formats, patient tongue positioning, physiological and/or environmental factors and patient tongue position variations. Test-retest analyses should also be conducted in future to validate the reliability of these property. An ML + DL hybrid model can significantly increase classification accuracy by combining the ability to learn from both ML and

DL algorithms. The system will therefore adopt the synergistic aspects of the classical ML algorithms and DL architectures to provide a more stable and generalizable result on a wider set of data.

References

1. Ting, S., & Zhang, B. (2014, December). Diabetes mellitus detection based on facial block texture features using the Gabor filter. In 2014 IEEE 17th International Conference on Computational Science and Engineering (pp. 1–6). IEEE.
2. Cibin, N. V., Franklin, S. W., & Nadu, T. (2015). Diagnosis of diabetes mellitus and NPDR in diabetic patient from tongue images using LCA classifier. Int J Adv Res Trends Eng Technol, 2, 57–62.
3. Sujatha, S., & Divya, D. (2015). A novel approach for analyzing diabetes mellitus and non-proliferative diabetic retinopathy using tongue. Indian Journal of Innovations and Developments, 4(2), 104.
4. Aher, S. A., & Dixit, V. V. (2015). Medical diagnosis using tongue colour analysis. International journal of Engineering, Education and Technology.
5. Lidia, T., & Ahmed, H. (2015). Tongue Image Analysis for Medical Diagnosis. International journal of scientific Engineering and Technology Research, 2263-2268.
6. Thirunavukkarasu, U., Umapathy, S., Ravi, V., & Alahmadi, T. J. (2024). Tongue image fusion and analysis of thermal and visible images in diabetes mellitus using machine learning techniques. Scientific Reports, 14(1), 14571.
7. Vocaturo, E., & Zumpano, E. (2020, December). Machine learning opportunities for automatic tongue diagnosis systems. In 2020 IEEE International Conference on Bioinformatics and Biomedicine (BIBM) (pp. 1498-1502). IEEE.
8. Festin, P. J. F., Cortez, R. S., & Villaverde, J. F. (2020, September). Non-Invasive Detection of Diabetes Mellitus by Tongue Diagnosis Using Convolutional Neural Network. In Proceedings of the 2020 10th International Conference on Biomedical Engineering and Technology (pp. 135–139).
9. Jiang, T., Hu, X. J., Yao, X. H., Tu, L. P., Huang, J. B., Ma, X. X., … & Xu, J. T. (2021). Tongue image quality assessment based on a deep convolutional neural network. BMC Medical Informatics and Decision Making, 21(1), 147.
10. Li.J., Chen, Q., Hu, X., Yuan, P., Cui, L., Tu, L.,… & Xu, J. (2021). Establishment of noninvasive diabetes risk prediction model based on tongue features and machine learning techniques. International Journal of Medical Informatics, 149, 104429.
11. Li, J., Yuan, P., Hu, X., Huang, J., Cui, L., Cui, J., … & Xu, J. (2021). A tongue features fusion approach to predicting prediabetes and diabetes with machine learning. Journal of biomedical informatics, 115, 103693.
12. Thanikachalam, V., Shanthi, S., Kalirajan, K., Abdel-Khalek, S., Omri, M., & Ladhar, L. M. (2022). Intelligent Deep Learning Based Disease Diagnosis Using Biomedical Tongue Images. Computers, Materials & Continua, 70(3).
13. Bharathi, M., Prasad, D., Venkatakrishnamoorthy, T., & Dharani, M. (2022). Diabetes diagnostic method based on tongue image classification using machine learning algorithms. Journal of Pharmaceutical Negative Results, 13(4), 1247-1250.
14. Li, J., Huang, J., Jiang, T., Tu, L., Cui, L., Cui, J., … & Xu, J. (2022). A multi-step approach for tongue image classification in patients with diabetes. Computers in Biology and Medicine, 149, 105935.
15. Bharani, R., Manishaa, K. T., Jaisurya, D., & Ram Kumar, C. (2023, March). Diabetes Mellitus Diagnosis based on Tongue Images using Machine Learning. In 2023 9th International Conference on Advanced Computing and Communication Systems (ICACCS) (Vol. 1, pp. 1614-1618). IEEE.

16. Zhang, N., Jiang, Z., Li, J., & Zhang, D. (2023). Multiple color representation and fusion for diabetes mellitus diagnosis based on back tongue images. Computers in Biology and Medicine, 155, 106652.
17. Hassoon, A. R., Al-Naji, A., Khalid, G. A., & Chahl, J. (2024). Tongue disease prediction based on machine learning algorithms. Technologies, 12(7), 97.

AI-Driven Precision Diagnostics for Sustainable Healthcare through Breast Cancer Detection Using HOA-Optimized Inception V3

U. Sabura Banu, M. Mohamed Ismail, and Ashutosh Mishra

Abstract Early detection of breast cancer is crucial to reducing global cancer mortality, a critical target under the United Nations Sustainable Development Goal 3 (SDG 3: Good Health and Well-being). This chapter introduces a novel artificial intelligence (AI) framework that leverages a deep convolutional neural network—Inception V3—optimized using the Hippopotamus Optimization Algorithm (HOA) to achieve high-performance breast cancer diagnosis from medical images. The method integrates deep learning and nature-inspired optimization to enhance diagnostic accuracy, reduce misclassification, and improve clinical usability. Comparative analysis across multiple CNN architectures (AlexNet, Xception, VGG16, ResNet50, and Inception V3) demonstrates that HOA-Inception V3 achieves superior performance with an accuracy of 97.13%, precision of 98.06%, and MCC of 0.9301, all while minimizing computational overhead. The research emphasizes the transformative potential of AI in sustainable healthcare systems and presents a replicable approach to intelligent medical diagnostics for other diseases and low-resource environments. This chapter contributes to SDG-aligned innovations by proposing intelligent, scalable, and resource-efficient AI models for early disease detection and healthcare equity.

Keywords Sustainable healthcare · SDG 3 · Deep learning · Breast cancer detection · Inception V3 · Hippopotamus Optimization Algorithm · Medical AI · Hyperparameter tuning

---

U. Sabura Banu (✉)
Department of Electrical and Electronics Engineering, Saveetha Engineering College, Chennai, Tamilnadu, India
e-mail: saburasec@gmail.com

M. Mohamed Ismail
B.S. Abdur Rahman Crescent Institute of Science and Technology, Vandalur, Chennai, Tamilnadu, India
e-mail: mmismail@crescent.education

A. Mishra
Department of EEE, Birla Institute of Technology and Science Pilani, Dubai Campus, Dubai, UAE

A. Khanna et al. (eds.), *Artificial Intelligence for Sustainable Development Goals*,
https://doi.org/10.1007/978-3-032-28217-0_9

1 Introduction

Breast cancer is the most prevalent cancer among women, accounting for a significant number of annual diagnoses and fatalities worldwide. The global burden of breast cancer remains a significant health challenge, particularly in low- and middle-income countries where diagnostic delays lead to poor prognosis. Considering reduction in lowering death rate of non-contagious illness, SDG 3 is addressed. Detecting the cancer with utmost care and higher accuracy at an early stage will improve the survival rate of the oncology patients. Medical intervention is required to detect the cancer from mammography, biopsy or histological data [1]. With the growing trend in deep learning techniques, various tasks such as image filtering, data augmentation, and classification of the medical images are gaining popularity [2, 3].

Various image processing such as filtering, segmentation, classification, and augmentation are performed using convolutional neural networks (CNNs) for several applications like agriculture, medical, satellite images, and industrial applications. Particularly, for medical applications, pretrained deep learning technique such as AlexNet, Xception, VGG16, Inception V3, and ResNet50 were proposed [4, 5], ensuring classification of benign and malignant medical images [6]. Even though, they show promising output in terms of efficiency and precision, they can be further optimized using heuristic algorithm [7]. Particularly, for the analysis of mammogram and detection of the breast cancer, integration of CNN and optimizer is showing a promising effect. Pretrained deep learning techniques are investigated by many researchers. Le et al. [8] proposed pretrained DL for classification of breast cancer. Similarly, Hussain et al. [9] demonstrated that VGG16 and ResNet50 were effective for the classification of breast cancer [9]. Usage of pretrained Inception V3 network assured that they can fit for image set from any domain. Among various DL techniques, Inception V3 outperforms in the case of medical imaging [10].

Even though the deep learning techniques show a promising outcome in medical image classification. The performance can be improved by optimizing the hyperparameters used in the pretrained networks. Even though mathematically proven optimization algorithms are available for centuries, in the past few decades, evolutionary and heuristic optimization techniques are applied in various domains. Various hyperparameters are optimized using heuristic algorithms, such as Particle Swarm Optimization (PSO) [11], Genetic Algorithms (GA) [12], and the Whale Optimization Algorithm (WOA) [13], thereby improving the performance of the already available deep learning techniques.

Out of the many heuristic optimization algorithms, Hippopotamus Optimization Algorithm (HOA) has a proven result for fine-tuning artificial intelligence models. It works on the basis of exploration and exploitation of the search space considering the global optimal solution [14]. Application of HOA for DL techniques is reported in the literature [15]. Integration of pretrained DL techniques with optimization is gaining importance. Sarker et al. [16] demonstrated the improved accuracy of the heuristic algorithm for image classification.

Zhang et al. [17] demonstrated the use of heuristic optimization for improved accuracy of pretrained deep learning technique. Chen et al. [18] suggested the inclusion of evolutionary algorithm-based deep learning for feature extraction. Shanthi et al. [19] worked on the detection of sub-types of breast cancer optimized transfer learning techniques over traditional algorithms for better accuracy. Shankar et al. [20] combined Extreme Learning Algorithm (ELA) with a deep feedforward network for classification of breast cancer. Kumar et al. [21] demonstrated the use of a hybrid DL model for the detection of breast cancer in which CNN was combined with SWM and random forests which showed better performance in terms of accuracy and reliability. In this proposed work, an attempt has been made to combine the benefits of fine tuning the hyper parameters of the Inception V3 model using HOA in terms of accuracy, recall, precision, etc.

Literature discussed in Table 1 clearly gives an insight into the various CNN techniques and optimizers for classification of medical images. Transfer learning and AI-driven solutions have been shown to improve classification accuracy by up to 96.2% in studies by Wang et al., Esteva et al., and Litjens et al. He et al. achieved 96.8% accuracy by enhancing gradient flow using residual learning in ResNet. Zhang et al. used data augmentation techniques to improve malignant tissue detection with 96.1% accuracy, whereas Hussain et al. concentrated on histopathology image classification using VGG16 and ResNet50, achieving 95.6%. The research on the application of optimization technique for the image classification is promising and has a bright future.

2 Materials and Methodology

2.1 Materials

Benign and malignant breast cancer images from the openly available repository were used in this study [22]. The image set consists of 620 benign and 250 malignant images. The data set is split into training (80%), validation (10%), and test data (10%).

2.2 Methodology

Several conventional deep learning techniques were applied. HOA-optimized Inception V3 is proposed. Comparison of the proposed optimal DL technique with the conventional DL technique is discussed in this section.

Table 1 Comparative analysis of optimization and interpretability techniques in deep learning-based breast cancer classification

Article/ Study	Methodologies Used	Technologies Used	Key Findings	Accuracy (%)	Contributions
Wang et al. [1]	Deep CNN for breast cancer detection	CNN, transfer learning	Improved classification accuracy	95.4	Enhanced feature extraction and classification
Esteva et al. [2]	AI-driven healthcare solutions	Deep learning, AI	Revolutionized medical diagnosis	94.8	AI applications in medical imaging
Litjens et al. [3]	Deep learning for medical image analysis	CNN, ResNet	High precision in classification	96.2	Comprehensive analysis of deep learning in healthcare
He et al. [4]	Residual learning for image recognition	ResNet, deep CNN	Improved gradient flow	96.8	Advanced residual learning techniques
Hussain et al. [6]	Breast cancer histopathological image classification	VGG16, ResNet50	High accuracy in classification	95.6	CNN models for histopathological images
Zhang et al. [7]	Deep learning for breast cancer imaging	CNN, data augmentation	Improved detection of cancerous tissues	96.1	Data augmentation strategies

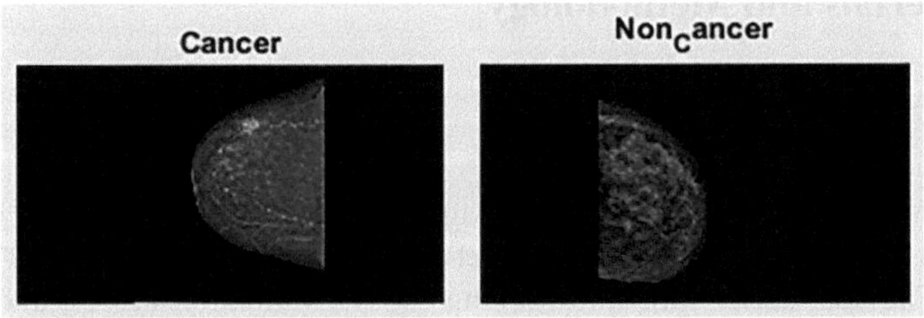

Fig. 1 Breast cancer oncology dataset sample

2.2.1 Data Preprocessing

The raw medical image needs to be preprocessed to ensure it is in proper condition to be used by the various CNN models. In addition, filtering of noise, improving the quality of the image by contrast, and brightness adjustment are also performed.

2.2.2 Model Selection

Few CNN techniques, such as AlexNet, Xception, VGG16, Inception V3, and ResNet50, were used for the comparison due to their proved ability for medical image classification.

AlexNet

AlexNet is the pioneer of the CNN technique used in image classification. The input to the AlexNet is a 227 x 227 x 3 size with 8 total layers. ReLU activation function was used with max pooling and local response normalization.

Xception

Xception is a CNN technique which is built on the Inception family where the inception modules are replaced by depthwise separable convolutions. Here, spatial and channel-wise features are learned separately. Xception works in two steps: depthwise convolution, which applies single filter to each channel separately, and pointwise convolution, which uses a 1 x 1 convolution for combining the depthwise convolution across channels, thereby reducing the number of parameters and computational time without compromising on the accuracy.

VGG16

VGG16 is a 16-layered deep convolutional neural network used for image classification with an input image size of 224 x 224 RGB. It has 13 convolutional layers and 3 fully connected layer, hence the name 16. Normally the activation function used is ReLU with Max pooling operation. Altogether, it has ~ 138 million parameters. Filters are 3x3 size consistently to enable capturing spatial hierarchies effectively.

Inception V3 (Default)

Several parallel convolutional filters with varying kernel sizes are incorporated into the sophisticated deep learning model Inception V3, which increases the effectiveness of feature extraction. To improve accuracy and lessen overfitting, the design makes use of auxiliary classifiers, factorized convolutions, and label smoothing. Because it can capture multi-scale spatial characteristics, Inception V3 is very good at identifying complicated medical images.

ResNet50

In order to address vanishing gradient problems in very deep networks, the deep residual network ResNet50 adds skip connections (residual learning). With the use of identity mappings, the 50-layer architecture enhances gradient flow during back-propagation. Because of its exceptional classification accuracy and resilience while working with massive image datasets, ResNet50 has been extensively used in medical image analysis.

2.2.3 Training of Models

The preprocessed dataset was used to train each model. The pretrained models were refined using the breast cancer dataset as part of the training process. Stochastic gradient descent (SGD) with momentum was used to optimize the models after they were trained using a cross-entropy loss function.

2.2.4 HOA Optimization

HOA was used to better enhance the Inception V3 model. Inspired by hippopotamus behavior, HOA is a metaheuristic optimization algorithm. It was utilized to enhance the Inception V3's performance by optimizing its hyperparameters.

Inception V3 Model for Breast Cancer Prediction

With deeper architectures and less computational complexity, the Inception V3 CNN is intended to improve classification performance. To increase efficiency, it makes use of methods like factorized convolutions.

- Asymmetric convolutions to improve the extraction of features.
- Batch normalization to enhance convergence; auxiliary classifiers to lessen overfitting have been pretrained and refined for the prediction of breast cancer.

2.2.5 Hippopotamus Optimization Algorithm (HOA) for Hyperparameter Tuning

Many conventional optimization techniques are in practice since centuries. The traditional mathematical optimization always suffers from getting stuck in local minima and many a times fail to reach the global optimal solutions. Heuristic algorithms and evolutionary optimization techniques reach out to the global optimal solutions due to the balance between exploration (global) and exploitation (local) to end up with the best solution. Hippopotamus Optimization Algorithm is one such heuristic population-based algorithm working with three steps: exploration, enabling the hippo to search for food in the entire search space; exploitation, where the search is focused in certain promising areas, and finally social interaction, where communication is done to share the information with the fellow hippos and enable decision-making. In the proposed technique, HOA-optimized Inception V3 is attempted to fine-tune the weight and bias learning rate to ensure optimized performance.

HOA-Optimized Inception V3 for Breast Cancer Prediction

Inception V3 DL model hyperparameters are optimized using the heuristic algorithm, HOA. The main purpose of the optimization is to increase accuracy, solve with decreased overfitting, improved computational cost, and faster convergence.

Step-by-Step HOA-Optimized Inception V3 Algorithm.

Step 1: Data Preprocessing

1. **Load Breast Cancer Dataset:**
   - Acquire benign and malignant breast cancer images from oncology repositories.
2. **Data Augmentation:**
   - Apply conventional image processing transformation to increase data diversity.
3. **Normalization:**
   - Scale pixel values to a range of **[0,1]** for faster convergence.
4. **Dataset Splitting:**
   - Divide data into **training, validation, and test sets (e.g., 80:10:10 split).**
5. **Convert Labels:**
   - Encode labels into categorical format for multi-class classification (e.g., **benign vs. malignant**).

Step 2: Initialize Inception V3 Model

1. **Load Pretrained Inception V3:**
   - Import a model pretrained on **ImageNet** and remove its top layers.
2. **Modify Output Layer:**
   - Replace the fully connected layer with a **Dense layer (Softmax or Sigmoid activation)** to match the number of breast cancer classes.

3. **Freeze Initial Layers:**
   - Keep early layers frozen to retain generic feature extraction capabilities.
4. **Define Trainable Layers:**
   - Select the deeper layers for fine-tuning based on hyperparameter optimization.

Step 3: Initialize HOA Parameters

1. **Define Population Size (N):**
   - Select an initial number of hippopotamus agents.
2. **Set Search Space:**
   - Define upper and lower bounds for hyperparameters:
     - Bias Learning Rate.
     - Weight Learning Rate.
3. **Random Initialization:**
   - Assign random values to each agent's hyperparameters.

Step 4: Evaluate Initial Fitness of Population

1. **Train the Inception V3 Model for a Few Epochs** with each agent's hyperparameters.
2. **Compute Fitness Score**
   - Use a weighted combination of performance metrics:
     $F = \alpha \times \text{Accuracy} + \beta \times \text{F1} - \text{score} - \gamma \times \text{Loss}$

- where α, β, γ are weight factors to prioritize accuracy, F1-score, and minimize loss.

3. **Store Best Initial Hyperparameters.**

Step 5: Update Hippopotamus Positions (Hyperparameter Tuning)

For each iteration (until convergence or max iterations):

1. Exploration Phase (Global Search)
   - Update agent positions using the global search equation:
     $X^{t+1} = X^{t} + r_1 \times (X_{best} - X^{t}) + r_2 \times D$
     , where
     - X^{t} is the current solution.

- X_{best} is the best-known solution.
- r_1, r_2 are random numbers between 0 and 1.
- D represents the random movement in the search space.

2. **Exploitation Phase (Local Search):**
   Refine promising solutions by adjusting step sizes: $X^{t+1} = X_{best} + \lambda \times (X^t - X_{best})$

   , where λ is a small adjustment factor.
3. Apply Constraints
   - Ensure hyperparameters stay within the predefined ranges.

Step 6: Train and Evaluate the Updated Inception V3 Model

1. **Use the Updated Hyperparameters** from HOA to fine-tune Inception V3.
2. **Train for a Fixed Number of Epochs** (e.g., 50–100) using the best agent's parameters.
3. **Validate the Model:**
   - Evaluate accuracy, loss, precision, recall, and F1-score on the validation set.
4. **Compare New Fitness Scores:**
   - If a new solution outperforms the previous best, update X_{best}.

Step 7: Check Convergence Criteria

1. **Stop Iteration if:**
   - The best fitness score does not improve for a predefined number of iterations.
   - Maximum iterations (e.g., 100) are reached.

Step 8: Final Model Testing and Performance Analysis

1. Evaluate the Optimized Model on the Test Set.
2. Compare Performance Metrics.

Step 9: Interpretability.

Explain Model Predictions Using Grad-CAM.

After applying HOA to optimize Inception V3 for breast cancer prediction, several improvements were observed. The model achieved higher accuracy, demonstrating better classification performance compared to default hyperparameters. Overfitting effect is a very important problem. Five-fold cross-validation is done for addressing the overfitting problem. In addition, fine-tuning of the conventional DL technique is done using the HOA algorithm to provide a balance between exploration and exploitation. Optimal parameter of the Inception technique guaranteed the faster convergence, improved efficiency, and performance. Overfitting was significantly reduced, ensuring a balanced generalization between the training and validation datasets. Faster convergence brought about by the optimization process also improved efficiency and performance by enabling the model to reach an optimal state in fewer epochs. The better performance suggests the use of the proposed heuristic optimized DL technique for medical applications in general.

2.2.6 Evaluation Metrics

Evaluation metrics are computed for the training and test data of the proposed model and the conventional techniques. Computational cost computation is done from the average inference time. Log Loss is also computed for understanding the effect of uncertainty.

2.2.7 Confusion Matrix

Confusion matrix gives a clear picture of the correctly classified and wrongly classified in an easily understandable manner.

3 Results and Discussion

Results of the proposed HOA-optimized Inception V3 and the traditional deep learning techniques are reported in this section. Even though it is a non-contagious disease, it is a life-threatening illness. Many DL techniques are used for the early detection of the illness. There is a growing demand for optimizing the hyperparameters of the conventional DL technique to improve the performance. An attempt is made to adopt various DL methods and improve the performance of the DL by fine-tuning the hyperparameters of the DL technique using heuristic algorithm; in this case, using Hippopotamus Optimization Algorithm, the results are reported in this section.

The confusion matrices in Fig. 2 illustrate the performance of six CNN models, displaying the correctly and incorrectly classified instances for the cancer and non-cancer categories. The matrices highlight true positives (correctly identified cancer cases), true negatives (correctly identified non-cancer cases), false positives (non-cancer cases misclassified as cancer), and false negatives (cancer cases misclassified as non-cancer). Notably, HPO-Inception V3 and Inception V3 achieved the best recall, with zero false negatives, ensuring all cancer cases were detected. ResNet50 had the highest true positive count (47), indicating strong cancer detection capability, while VGG16 also performed well, misclassifying only 6 cancer cases as non-cancer. However, Xception and AlexNet exhibited relatively higher false negatives, suggesting they missed more cancer cases. Overall, HPO-Inception V3 and ResNet50 emerged as the most reliable models, offering high accuracy with minimal misclassification.

The results of the study shown in Table 2 are summarized. The key findings are as follows:

- Accuracy: HOA-Inception V3 achieved the highest accuracy of 97.13%; next is ResNet50 with 94.83% accuracy.

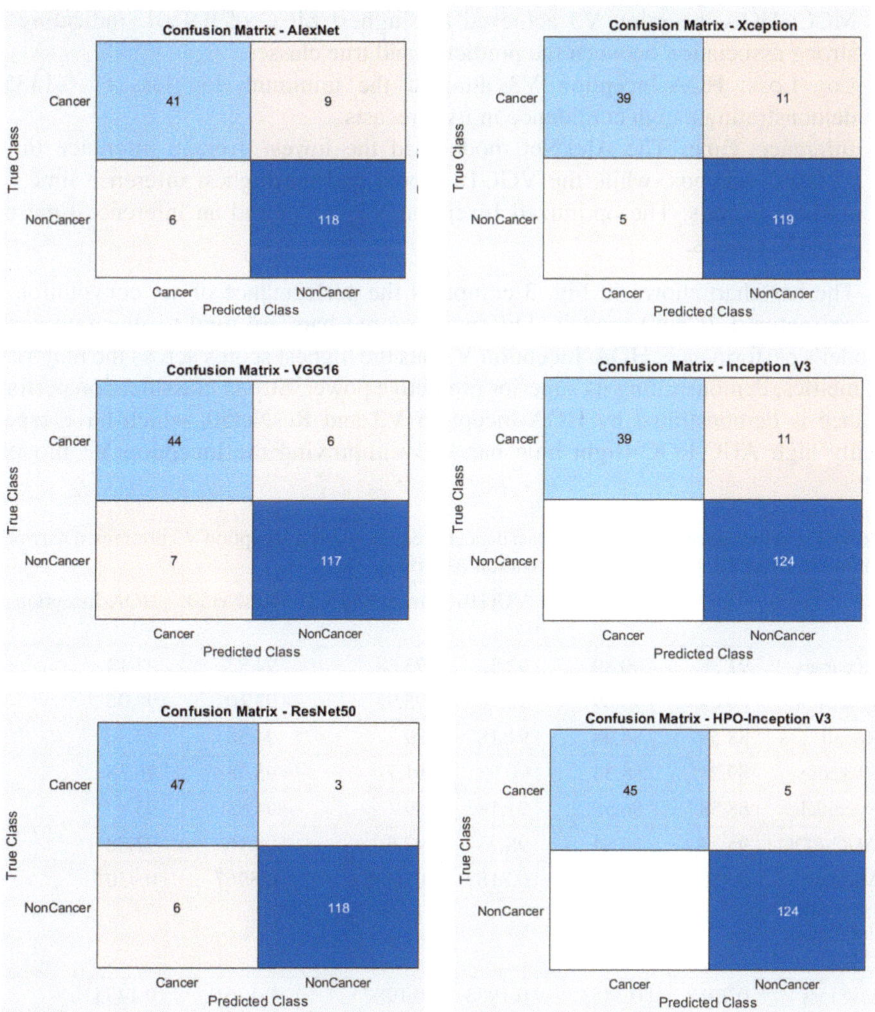

Fig. 2 Confusion matrices of six CNN models for breast cancer prediction

- Precision: The optimized Inception V3 model also gave the best precision of 98.06%, demonstrating a very low rate of false positives.
- Recall: The ResNet50 model had the highest recall of 94.58%, while HOA-Inception achieved a recall of 95%.
- F1-score: HOA-Inception method showed the high F1-score of 96.38%, showcasing equilibrium between precision and recall.
- Specificity: Specificity of ResNet50 is 94.58%, while the HOA-Inception V3 model achieved a specificity of 95%.
- AUC-ROC: AUC-ROC score of VGG16 model is 98.73%, followed closely by the optimized Inception V3 model with 97.82%.
- MCC: HOA-Inception V3 achieved the highest MCC of 0.9301, indicating a strong association between the predicted and true classes.
- Log Loss: HOA-Inception V3 attained the minimum log loss of 0.1435, demonstrating a high confidence in its forecasts.
- Inference Time: The AlexNet model had the lowest average inference time of 0.067 seconds, while the VGG16 model had the highest inference time of 0.9546 seconds. The optimized Inception V3 model had an inference time of 0.2415 seconds.

The bar chart shown in Fig. 3 compares the performance of six convolutional neural network (CNN) models. Different colored bars are used to illustrate each model's performance; HOA-Inception V3 has the highest scores across the majority of metrics, demonstrating its superior predictive power. Strong classification performance is demonstrated by HOA-Inception V3 and ResNet50, which have especially high AUC-ROC (light blue bars). By improving the Inception V3 model,

Table 2 Performance matrices for breast cancer prediction using Inception V3 pretrained network optimized using Hippopotamus Optimization Algorithm

	AlexNet	Xception	VGG16	Inception V3	ResNet50	HOA-Inception V3
Accuracy	91.38	90.80	92.53	93.68	94.83	97.13
Precision	90.07	90.09	90.70	95.93	93.10	98.06
Recall	88.58	86.98	91.18	89	94.58	95
F1-score	89.28	88.34	90.93	91.7	93.78	96.38
Specificity	88.58	86.98	91.18	89	94.58	95
AUC-ROC	95.77	94.84	98.73	97.6	98.16	97.82
Matthews correlation coefficient (MCC)	0.7864	0.7701	0.8187	0.0846	0.8767	0.9301
Log loss	0.2015	0.3453	0.1953	0.1987	0.1761	0.1435
Average inference time (sec)	0.067	0.667	0.9546	0.3149	0.1359	0.2415

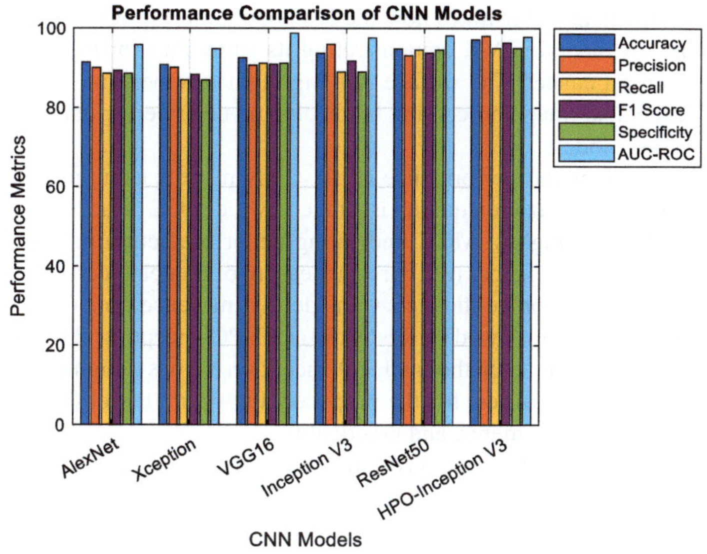

Fig. 3 Bar chart for performance comparison of CNN models for breast cancer prediction

the Hippopotamus Optimization Algorithm (HOA) makes it the most accurate and dependable model for detecting breast cancer, as seen in Fig. 3.

A comparison depiction of the performance of various CNN models is provided by the 3D surface plot displayed in Fig. 4. Variations are represented by the plot's color gradient.

From AlexNet to HOA-Inception V3, the trend shows a steady improvement in performance, underscoring the value of optimizing deep learning models in Fig. 4.

Fig. 4 CNN performance metrics for breast cancer prediction in a 3D surface plot

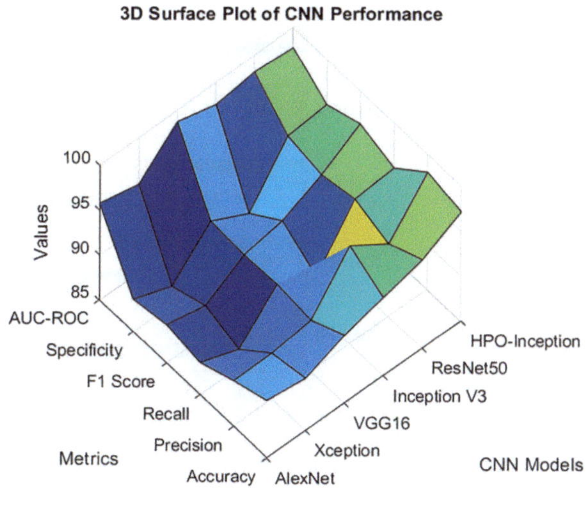

This 3D graphic clearly illustrates how the Hippopotamus Optimization Algorithm (HOA) improves Inception V3, demonstrating its suitability for the most accurate detection of breast cancer. The surface plot highlights the benefits of optimization strategies in medical image processing by displaying clear performance differences between various models.

Figure 5 examines the performance of six CNN models using a horizontal bar chart. The entire categorization quality is gauged by the MCC (blue bars), and HOA-Inception V3 has the highest MCC, indicating better prediction performance. For HOA-Inception V3, the log loss (red bars), a measure of model uncertainty, is lowest, indicating more prediction confidence. Considering inference time (yellow bars), it is minimal for AlexNet, followed by ResNet50 and HPO-Inception V3. On analyzing the difference algorithms, even though HPO-Inception V3 has a slightly higher inference time, the accuracy is best for the proposed algorithm, thereby making it best suitable in terms of performance and practicality.

The 3D surface plot which analyzes and compares the proposed algorithm with its traditional conventional deep learning technique counterparts with MCC, Log Loss, and Inference Time. Performance is guaranteed using the highest MCC of the proposed technique which shows improved detection trustworthiness. Considering the Log Loss, it is minimum for the proposed HOA-Inception V3 model and ResNet, proving that the predictions are accurate even under uncertain conditions. Analysis of inference time of various models clearly demonstrates that VGG16 and Inception V3 have the maximum time compared to the proposed HPO optimized Inception V3, proving that the proposed algorithm has a good balance between computation economy and accuracy.

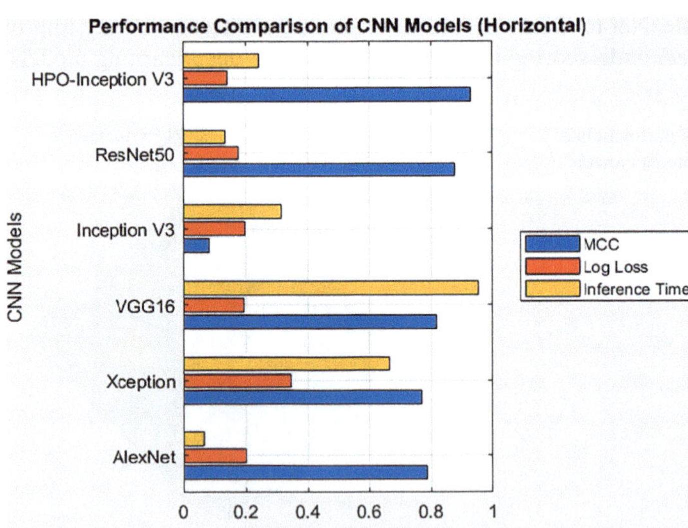

Fig. 5 Performance comparison of CNN models based on MCC, Log Loss, and Inference Time

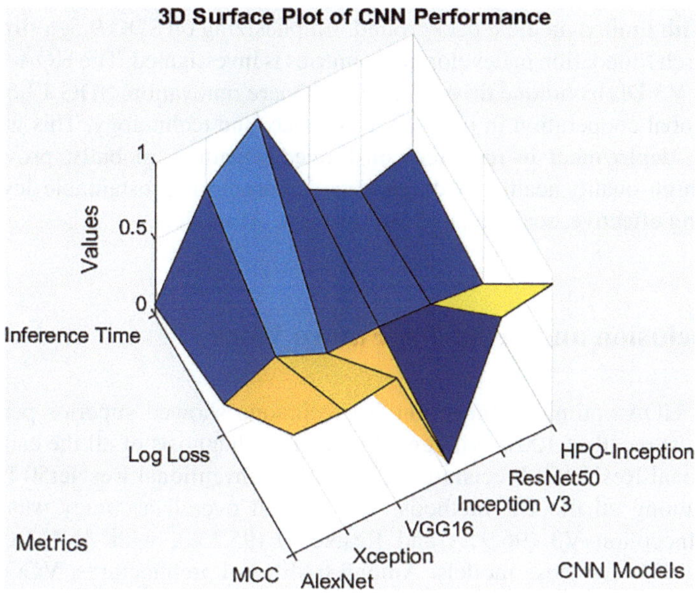

Fig. 6 3D Surface plot of CNN performance based on MCC, Log Loss, and Inference Time

Overall finding of the proposed algorithms across various performance metrics shows that the HOA-Optimized Inception V3 model outperforms other conventional DL techniques adopted. Inception V3 model hyperparameters optimized by the HOA heuristic algorithm show enhanced performance. It differentiates between the cancerous and benign cells with a much higher accuracy, high precision, and recall scores. Modest log loss guarantees the model's confidence. Even though VGG16 model outperforms in identifying between the two groups, it is not practically adaptable due to its longer inference time. Considering ResNet50, it has a good recall and is applicable for low false-negative cases. Even though AlexNet has the shortest inference time, it has the lowest accuracy and F1-score. AlexNet is not recommended due to its low precision and accuracy. Overall, the proposed optimized Inception V3 model shows exceptional performance in terms of accuracy, precision, recall, and computational cost.

4 Sustainability Alignment of the Proposed Technique

The proposed technique tends to address various SDGs, thereby helping in early identification of breast cancer and improving the survival rate of oncology patients, emphasizing SDG 3.4. It aligns with SDG 3.d in providing an early warning system and risk reduction for health solutions. Deep learning techniques provide early diagnosis, helping healthcare providers deliver better clinical assistance in a timely manner with limited medical background. Emphasizing on SDG 9.5, a strong scientific research foundation in developing countries is investigated. The HOA-optimized Inception V3 DL technique drives AI in healthcare innovation. SDG 17.6 is met by way of global cooperation in innovation, science, and technology. This work offers promising deployment in resource-constrained situations globally, providing fair access to high-quality healthcare diagnostics and promoting sustainable development by enabling effective, scalable, and interpretable AI models.

5 Conclusion and Scope for Future Work

Proposed HOA-optimized Inception V3 technique showed superior performance with highest recall of 100% with ensuring correct diagnosis of all the cancer cases. Conventional ResNet50 Precision (94%) of the conventional ResNet50 is 94% the highest among all the DL methods. The highest overall accuracy was observed in HPO-Inception V3 (96.9%) and ResNet50 (95.3%), with AUC-ROC scores exceeding 98% for these models. Among traditional architectures, VGG16 exhibited balanced performance with an accuracy of 92.8%, while AlexNet and Xception lagged with lower recall and specificity. The confusion matrix analysis confirmed that HPO-Inception V3 and Inception V3 eliminated false negatives, making them strong candidates for medical diagnostics. However, inference time analysis revealed that VGG16 and Inception V3 required significantly more computational resources, making them less efficient for real-time applications. The approach is not limited to breast cancer but can be generalized across disease areas, supporting broader healthcare-related SDGs. For future improvements, reducing inference time while maintaining classification accuracy is a priority. Model optimization techniques such as pruning, quantization, and hardware acceleration can significantly enhance efficiency. Additionally, hybrid deep learning approaches and transfer learning on larger, diverse datasets can improve robustness and generalizability.

References

1. S. Wang, J. Yu, D. Kang, R. Wang, J. Huang, and Y. Zheng, "Breast cancer detection using deep learning techniques: A survey," *IEEE Access*, vol. 8, pp. 220891-220910, 2020.
2. B. Esteva, A. Kuprel, R. A. Novoa, J. Ko, S. M. Swetter, H. M. Blau, and S. Thrun, "Deep learning for healthcare: Review and perspectives," *Nature Medicine*, vol. 25, no. 1, pp. 24-29, 2019.
3. G. Litjens, T. Kooi, B. Ehteshami Bejnordi, A. A. A. Setio, F. Ciompi, M. Ghafoorian, J. A. W. M. Van Der Laak, B. Van Ginneken, and C. I. Sánchez, "A survey on deep learning in medical image analysis," *Medical Image Analysis*, vol. 42, pp. 60-88, 2017.
4. K. He, X. Zhang, S. Ren, and J. Sun, "Deep residual learning for image recognition," in *Proc. IEEE Conf. Comput. Vis. Pattern Recognit. (CVPR)*, 2016, pp. 770-778.

5. C. Szegedy, V. Vanhoucke, S. Ioffe, J. Shlens, and Z. Wojna, "Rethinking the Inception architecture for computer vision," in *Proc. IEEE Conf. Comput. Vis. Pattern Recognit. (CVPR)*, 2016, pp. 2818-2826.
6. Y. LeCun, Y. Bengio, and G. Hinton, "Deep learning," *Nature*, vol. 521, no. 7553, pp. 436-444, 2015.
7. S. J. Pan and Q. Yang, "A survey on transfer learning," *IEEE Trans. Knowl. Data Eng.*, vol. 22, no. 10, pp. 1345-1359, 2010.
8. T. Le, M. Materka, M. K. Islam, P. Saripan, A. F. Mohamed, and N. Salleh, "Transfer learning for breast cancer classification in mammograms," *IEEE Access*, vol. 9, pp. 13746-13755, 2021.
9. M. Hussain, S. L. M. S. Hameed, R. Alzahrani, M. Fatima, and A. Qamar, "Breast cancer histopathological image classification using deep learning," *Future Gener. Comput. Syst.*, vol. 98, pp. 346-356, 2019.
10. C. Szegedy, W. Liu, Y. Jia, P. Sermanet, S. Reed, D. Anguelov, D. Erhan, V. Vanhoucke, and A. Rabinovich, "Going deeper with convolutions," in *Proc. IEEE Conf. Comput. Vis. Pattern Recognit. (CVPR)*, 2015, pp. 1-9.
11. J. Kennedy and R. Eberhart, "Particle swarm optimization," in *Proc. IEEE Int. Conf. Neural Netw.*, 1995, pp. 1942-1948.
12. J. H. Holland, *Adaptation in Natural and Artificial Systems*, MIT Press, 1975.
13. S. Mirjalili and A. Lewis, "The whale optimization algorithm," *Adv. Eng. Softw.*, vol. 95, pp. 51-67, 2016.
14. R. Kumar, B. Singh, A. K. Gupta, and R. Pandey, "Hippopotamus optimization algorithm: A novel metaheuristic approach," *Appl. Soft Comput.*, vol. 113, 2021.
15. Y. Zhang, L. Sun, Y. Du, and H. Wang, "Optimized deep CNN for medical image classification using metaheuristic algorithms," *IEEE Access*, vol. 9, pp. 22365-22375, 2021.
16. M. Sarker, R. H. Haque, T. Rahman, S. Das, and K. Dey, "Metaheuristic optimization in medical diagnosis," *IEEE Access*, 2022.
17. Y. Zhang, W. Liu, X. Zhao, and P. Huang, "Deep learning in breast cancer imaging," *Nat. Sci. Rep.*, 2023.
18. X. Chen, H. Li, C. Wang, and T. Zhang, "Evolutionary algorithms for medical image analysis," *IEEE Trans. Med. Imaging*, 2023.
19. D. Shanthi, K. Rajalakshmi, and G. Santhi, "A study of deep learning techniques for predicting breast cancer types," *Soft Computing*, vol. 27, pp. 14532-14549, 2023. DOI:https://doi.org/10.1007/s00500-023-08245-2.
20. G. S. Shankar, M. Priya, R. Balasubramanian, and A. Karthikeyan, "Breast cancer diagnosis using virtualization and extreme learning algorithm based on deep feed forward networks," *Biomed. Eng. Comput. Biol.*, vol. 15, pp. 1-12, 2024. DOI:https://doi.org/10.1177/11795972231183999.
21. R. Kumar, P. S. Kumar, and B. Vijayakumar, "Hybrid deep learning model for automated breast cancer detection in histopathological images," *Multimedia Tools and Applications*, vol. 82, no. 12, pp. 15679-15697, 2023. DOI:https://doi.org/10.1007/s11042-023-14451-6.
22. K. B. Aqdar, P. A. Abdalla, R. K. Mustafa, Z. H. Abdulqadir, A. M. Qadir, A. A. Shali, and N. M. Aziz, "Mammogram Mastery: A Robust Dataset for Breast Cancer Detection and Medical Education," *Mendeley Data*, V1, 2024. DOI: 10.17632/fvjhtskg93.1.

Accessing Olaichuvadi Using Artificial Intelligence for Establishing Sustainability in Preserving Cultural Heritage

U. Sabura Banu, M. Mohamed Ismail, and Abdul Rajak A. R.

Abstract In modern world, there is a need to preserve cultural heritage preservation to ensure sustainable development (SDG 11.4). Olaichuvadi in Tamilnadu is one such cultural wealth where abundance of information wealth is available. Olaichuvadi is our ancient Tamil literature written in palm leaves written before centuries and preserved in central libraries and universities. Even though it is preserved, many a times it undergoes dryness, damaged by pests, affected by humidity or damaged by physical handling. There is a need to digitize the olaichuvadi to ensure they are preserved for future generations. To retrieve and for storage of the content of the olaichuvadi, binarization techniques are required. In addition to retrieval and storage, binarization enhances the quality of the retrieved manuscript for interpretation. In this study, various thresholding techniques including global thresholding and adaptive binarization techniques are attempted to get the information from the olaichuvadi. But the conventional thresholding techniques often face challenges due to the uneven illumination, noise, and text degradation. In this article, various conventional binarization techniques including Wolf Binarization, Niblack Thresholding, Sauvola Thresholding, Bernsen Thresholding, Nick Thresholding, and Bradley Thresholding are attempted. Even though the above-said binarization techniques can be adapted for digitization of the olaichuvadi, mostly they face problems due to the external noise and illumination and contrast irregularity, lack of adaptation to complicated degraded palm scripts, etc. To optimize their performance, there is a need to tune the parameters of the conventional techniques. In this article, an attempt has been made to optimize

U. Sabura Banu (✉)
Department of Electrical and Electronics Engineering, Saveetha Engineering College, Chennai, Tamilnadu, India
e-mail: saburasec@gmail.com

M. Mohamed Ismail
B.S. Abdur Rahman Crescent Institute of Science and Technology, Vandalur, Chennai, Tamilnadu, India
e-mail: mmismail@crescent.education

A. Rajak A. R.
Department of EEE, Birla Institute of Technology and Science Pilani, Dubai Campus, Dubai, UAE

A. Khanna et al. (eds.), *Artificial Intelligence for Sustainable Development Goals*,
https://doi.org/10.1007/978-3-032-28217-0_10

the parameters using Whale Optimization Algorithm (WOA) and Wild Horse Optimization (WHO). Comparison is made to show the superiority of the proposed technique with the conventional techniques. The experimental results reveal that while Wild Horse Optimization (WHO) demonstrated strong performance with a Precision of 0.83556, Recall of 0.58507, and F-measure of 68.82356%, the Whale Optimization Algorithm (WOA) ultimately proved to be more efficient. WOA achieved a higher Precision of 0.8668 and a superior overall F-measure of 81.469%, along with a Geometric Accuracy of 0.75929 and Balanced Classification Rate (BCR) of 78.522%. Among the adaptive binarization techniques, the highest Precision 0.79637 was displayed by Bradley Thresholding, while Nick Thresholding performed best in BCR and PSNR. However, the Best Scores delivered by WOA were consistently the best (0.0745) and its PSNR superior 12.2370; therefore, WOA is the ideal choice for clearing up the image and distortion in manuscript restoration. These findings highlight the efficiency and effectiveness of the Whale Optimization Algorithm in the binarization process for preserving ancient manuscripts. Olaichuvadi stands as a testimony to how AI can bridge the digital divide in heritage preservation, ensuring equitable access to indigenous knowledge for future generations while contributing to SDG 4 (Quality Education), SDG 9 (Innovation and Infrastructure), and SDG 17 (Partnerships for Goals).

Keywords Palm leaf manuscripts · Cultural heritage AI · SDG 11.4 · Semantic OCR · Binarization · Tamil manuscripts · Olaichuvadi · Sustainable education · Digital libraries

1 Introduction

Olaichuvadi is a treasured documentation version of the Tamil culture with abundant information in diverse fields such as culture, medicine, and history. Preserving the olaichuvadi for centuries is a mind boggling challenge posed to the present generation. In the ever-increasing growth of computer vision and artificial intelligence, there is path opened for digital preservation of the information. Both digitization and restoration of the information can be done using digital techniques. In addition to restoration and storage for future generation, the content in the olaichuvadi should be made readable. Uneven lighting effects, faded or noise in the palm leaves make the text unreadable. There is need to binarize the content and make it readable. Binarization is a technique used to convert the gray scale image to a black and white image. Various conventional binarization techniques are adopted for the binarization. Most of the traditional techniques faces problem due to the problems with the lighting effects, contrast adjustment, noise, etc. Fixing a single threshold value across the entire image provides poor output. Adaptive binarization techniques have emerged, offering localized thresholding based on the intensity distribution of pixels within a defined neighborhood. Techniques such as Wolf Binarization, Niblack Thresholding, Sauvola Thresholding, Bernsen Thresholding, Nick Thresholding, and

Bradley Thresholding have been developed to enhance image clarity by dynamically adjusting the threshold values according to local pixel characteristics.

These methods seem promising but need further improvement so that they can provide the best possible results for binarization, especially in highly degraded and noisy images. Metaheuristic algorithms such as WOA and WHO have been recently of immense interest because they can optimize complex functions with ease. These algorithms can be used to tune parameters for binarization methods so that better restoration results can be achieved.

In the proposed olaichuvadi preservation, WOA and WHO optimization techniques are used for tuning the parameters of the thresholding techniques. Global optimal threshold value is obtained using the heuristic optimization techniques. Final objective is to get an improved readability of the text by minimizing the fitness function comprising binarization metrics such as Precision, Recall, and F-measure. Comparison of the proposed optimized thresholding techniques with the conventional techniques is done. On investigation of the results, it is evident that WHO-optimized adaptive binarization ensures superior digitization and restoration quality, thereby preserving the traditional and historical documents. The process of preservation in digital form and restoration of the content aligns with the United Nations Sustainable Development Goals (SDGs) by providing a means of restoration of the olaichuvadi and make it accessible to the future generations ensuring sustainable knowledge transmission through intelligent technologies.

2 Literature Survey

Many literatures have discussed the binarization of the palm scripts in India, China, Nepal, Sri Lanka, Burma, Thailand, Indonesia. Mostly the literatures focus on various conventional binarization techniques. Optimization of the parameters of the thresholding techniques is also reported in literature. Alexander and Kumar [1] discusses optimization of WOA-tuned binarization technique for preserving the content of the olaichuvadi. They could achieve clarity in degraded olaichuvadi. The other is from Poornima Devi and Sornam [2] where feature engineering techniques are introduced to improve text extraction and classification of Tal-Patra manuscripts, based on natural language processing, [2]. Historical preservation methods were analyzed by Wiland et al. [3], who provided a comprehensive overview of palm leaf manuscript conservation practices, from the preparation of the leaves to addressing the common types of damage encountered [3]. Nishanthi [4] discussed various indigenous methods to restore olaichuvadi [4].

Chu et al. [5] focused on analyzing the Burma palm scripts degradation due to physical handling or environmental effects [5]. Lertratanakehakarn [6] deals with restoration of the Thailand palm scripts highlighting the role of digital archives for digitization of the manuscripts for future [6]. Michalak and Okarma [7] dealt with binarization in case of varied illumination affecting the binarization of affected manuscripts [7]. Michalak et al. [8] applied Generalized Gaussian Distributions

(GGD) and Gaussian Mixture Models (GMM) for upgrading the restoration and readability of papyri images [8]. Bloechle et al. [9] introduced ZigZag algorithm which worked for varied illuminated palm script for binarization of ancient texts [9].

Lins et al. [10] evaluated efficiency by assessment of image quality, processing time, and size of the photographed documents for restoration of ancient text [10]. Amiriebrahimabadi and Mansouri [11] discussed Whale-optimized feature selection techniques for improved efficiency in various tasks, including manuscript restoration [11]. Lin [12] dealt with upgraded version of WOA for Kernel Extreme Learning Machine for classification and binarization [12].

Zheng et al. [13] discussed about enhanced variant of the WHO outperforming WOA for computer vision and ancient text restoration problems [13]. Ju et al. [14] dealt with textural images using three-stage binarization method for color document images using Discrete Wavelet Transform (DWT) and Generative Adversarial Networks (GANs) [14].

Deep learning-based foreground and background segmentation of Lanna Archaic for improving the digitization and readability of the ancient scripts was dealt by Pravesjit and Seng [15]. Subramani and Subramaniam [16] dealt with generating olaichuvadi dataset for Tamil character recognition contributing to the preservation of traditional Tamil texts [16].

Paulus et al. [17] used extraction technique for line retrieval from olaichuvadi for the analysis of text document [17]. Sudarsan and Sankar [18] worked on Malayalam palm leaf manuscripts in terms of noise reduction and improving restoration of degraded documents with advanced computer vision techniques [18]. Haritha et al. [19] focused on recognition of characters in palm scripts, and Thuon et al. [20] utilized CNNs and glyph classification for restoration of characters of ancient palm manuscripts [19, 20].

Even though there are lot of works reported in the binarization and restoration of the palm scripts, there are few research gaps to be addressed as reviewed in literature. Merging traditional conservation methods with cutting-edge technologies such as AI, deep learning, and optimization algorithms provides immense opportunities for developing precision and effectiveness further in restoration and digitization processes. Future research is likely to focus on improving hybrid techniques, developing classification models, and enlarging datasets for more extensive regional and ancient scripts.

3 Methodology for Palm Leaf Manuscript Using Various Binarization Techniques

In the preservation of manuscripts, effective image binarization technologies are to be employed to enhance text clarity and readability while preserving the underlying structure of the manuscripts. The following methodology describes how

Fig. 1 Original image of olaichuvadi (palm leaf)

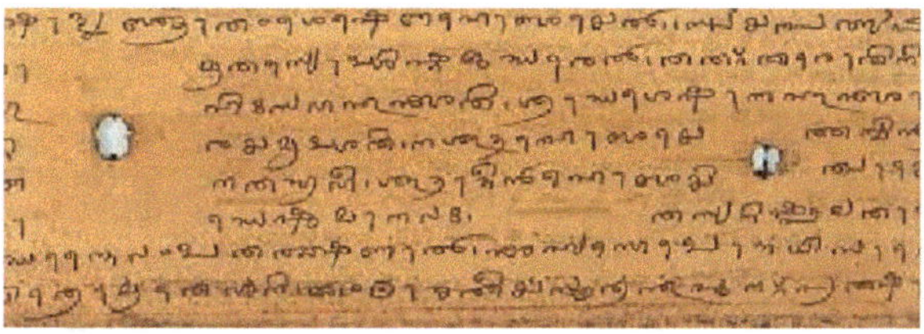

different binarization techniques including Wolf, Nick, Bradley, Niblack, Sauvola, and Bernsen methods are applied.

3.1 Image Acquisition and Preprocessing

High-resolution images of palm leaf manuscripts were taken using advanced scanners or cameras for minimal image distortion. Removing noise and enhancing quality prior to binarization were preprocessing tasks that were performed.

3.2 Binarization Techniques

Each binarization technique makes use of a distinct mathematical model that separates the text from the background in the conversion of a grayscale image to a binary (black and white) image. We apply the following binarization methods. The original image of olaichuvadi (palm leaf) is shown in Fig. 1.

3.3 Wolf Binarization

Wolf binarization adapts the thresholding technique to manage unevenly illuminated images. It develops a local threshold based on the local mean and standard deviation of pixel intensities that is derived from Eq. 1. Wolf Binarization is especially applicable to documents receiving non-uniform illumination such as degraded palm leaf manuscripts that are likely to present varying lighting conditions over the surface is shown in Fig. 2

$$T(x, y) = m(x, y) + k.v(x, y) - m(x, y).\frac{I(x.y) - m(x, y)}{v_{\max} - m_{\min}} \tag{1}$$

In this equation, $T(x, y)$ is the threshold value at the pixel (x, y), $m(x, y)$ is the local mean of the neighborhood pixel (x, y), is the standard deviation, $v_{\max}$ is the

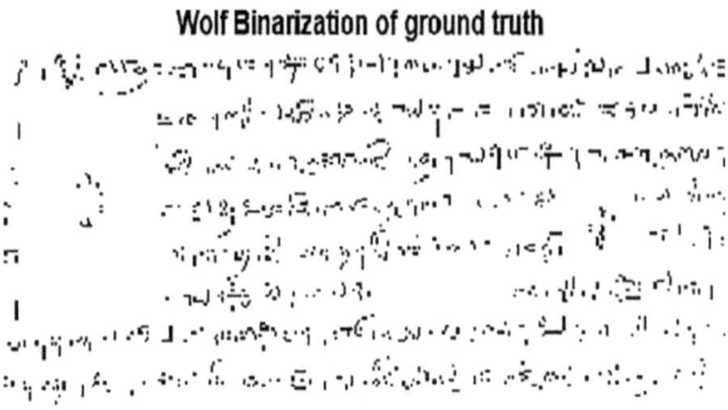

Fig. 2 Wolf binarization of olaichuvadi (Palm leaf manuscript)

minimum mean across the entire image, mmin is the minimum mean across the entire image, and k is a constant which allows adjustment of the threshold. $I(x, y)$ of the neighborhood pixel, (x, y) is the standard deviation, v_{max} is the minimum mean across the entire image, m_{min} is the minimum mean across the entire image, and k is a constant which allows adjustment of the threshold.

3.4 Nick Binarization

Nick binarization is an adaptive thresholding technique with high-speed performance specifically designed for binarization of document images. It is shown in Fig. 3. It works on the idea of using local mean values for determining threshold values on each pixel. It's derived from Eq. 2. Nick binarization is computationally efficient and ideal for huge collections of manuscript images when high-speed processing is imperative.

$$T(x, y) = m(x, y) - k\sqrt{\frac{v(x, y)}{1 + s(x, y)}} \tag{2}$$

$T(x, y)$ is the local threshold; $m(x, y)$ is the local mean of the neighborhood pixels; $v(x, y)$ is the variance of the pixels; $s(x, y)$ is the smoothing term to avoid division by zero; k is a constant factor.

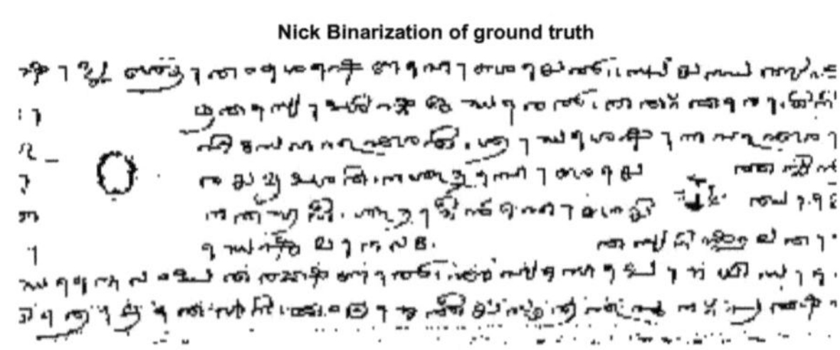

Fig. 3 Nick binarization of olaichuvadi (palm leaf manuscript)

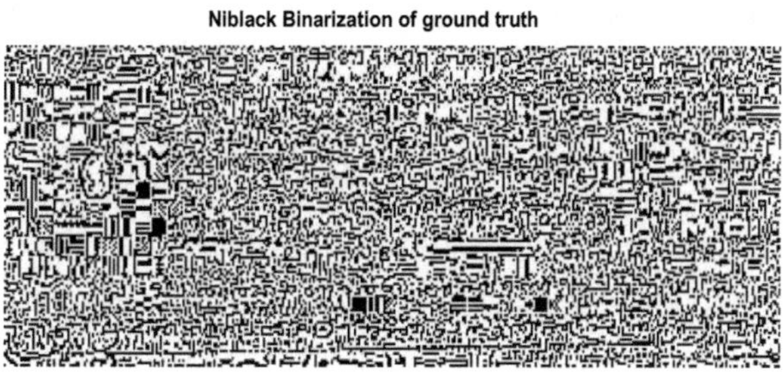

Fig. 4 Niblack thresholding of olaichuvadi (palm leaf manuscript)

3.5 Niblack Thresholding

Niblack thresholding works by considering the local mean and standard deviation of pixel values within a sliding window of size, which is shown in Fig. 4. It adjusts the threshold value based on the local pixel distribution. Niblack is well-suited for areas of texts on manuscripts with varying densities and degraded locations, as the value of the threshold depends on local variations. The threshold is calculated by Eq. 3.

$$T(x, y) = m(x, y) + k.v(x, y) \tag{3}$$

$T(x, y)$ is the threshold at pixel (x, y), $m(x, y)$ is the local mean of the neighborhood pixels, $v(x, y)$ is the local standard deviation, k is a constant factor between (− 0.2 and − 0.5).

3.6 Sauvola Thresholding

Sauvola thresholding is an improved method of Niblack with a special focus on noise reduction and maximum effectiveness on heavily degraded manuscripts. It takes into account the local mean and standard deviation but adds some additional weight. This is shown in Fig. 5. Sauvola is effective for documents of any kind of noise or damage. This is particularly useful in preserving the fine details of palm leaf manuscripts. Sauvola thresholding can now be calculated by Eq. 4.

$$T(x, y) = m(x, y)(1 + k\left(\frac{v(x, y)}{R} - 1\right) \tag{4}$$

$T(x, y)$ is the threshold at pixel (x, y), $m(x, y)$ is the local mean of the neighborhood pixels, $v(x, y)$ is the local standard deviation, k is a constant factor, R is the dynamic range of standard deviation, typically set to 128.

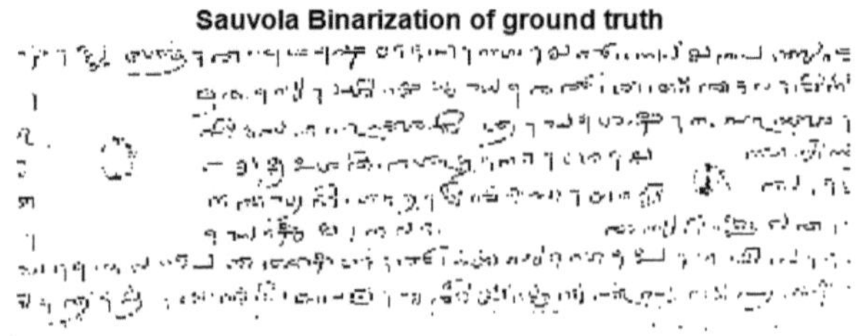

Fig. 5 Sauvola thresholding of olaichuvadi (palm leaf manuscript)

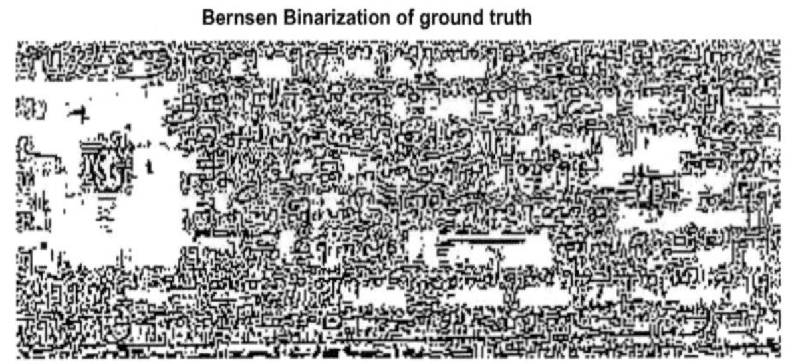

Fig. 6 Bernsen thresholding of olaichuvadi (palm leaf manuscript)

3.7 Bernsen Thresholding

Bernsen thresholding may be defined as a contrast algorithm that relies on Eq. 5. It calculates the threshold from the average contrast taken from the minimum and maximum pixel values, in one local window, while writing by developing the threshold. Bernsen's algorithm is relatively useful where said contrast exists between text and background areas of the manuscripts and is shown in Fig. 6, and by its virtue, it adds maximum contrast to the text.

$$T(x, y) = \frac{I_{\max}(x.y) + I_{\min}(x, y)}{2} \tag{5}$$

$T(x, y)$ is the threshold at pixel (x, y), $I_{\max}(x.y)$, $I_{\min}(x.y)$ are the maximum and minimum pixel intensities in a local window around(x, y).

3.8 Bradley Binarization

Bradley binarization is a thresholding technique where the threshold is computed based on a fixed percentage of the average pixel values within a local window encircling each pixel given by Eq. 6. Bradley's method is very effective for palm leaf manuscripts with faint text and uneven lighting, allowing it to dynamically adjust the threshold in different areas, is shown in Fig. 7.

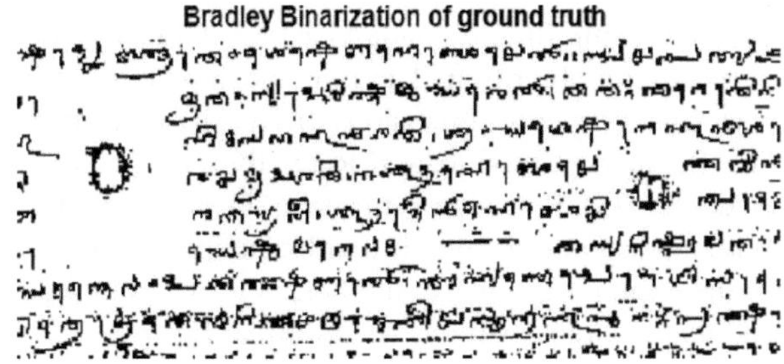

Fig. 7 Bradley binarization of olaichuvadi (palm leaf manuscript)

$$T(x, y) = \frac{1}{N} \sum_{(i,j \in W)} I(i, j)(1 - p) \tag{6}$$

$T(x,y)$ is the threshold for pixel(x, y), N is the number of pixels in the local window W, $I(i, j)$ is the pixel intensity at location (i, j), p is the constant percentage (usually around 0.15), and k is a constant factor.

3.9 *Post-Binarization Processing*

Binarized images that result are processed for further segmentation, character recognition, and connotation processes. These digital files are stored for archival. Each binarization technique is assessed using precision, accuracy of detected text pixels, and recall ability to capture faint text. PSNR Peak signal-to-noise ratio for evaluating image quality. SSIM Structural similarity index for comparing binary images to the original grayscale images.

4 Wild Horse Optimization (WHO) Algorithm with Bradley Binarization

In order to devise a way to effectively perform image binarization, especially for the preservation of inscriptions on palm leaves, the scheme proposes a combination of the Wild Horse Optimization (WHO) algorithm and Bradley Binarization to effectively set aside text from the background. The WHO algorithm optimizes the threshold and window size for Bradley's adaptive binarization, and the binarization results are evaluated on several performance metrics including Precision, Recall, F-measure, Sensitivity, between two epidemiological specificities.

Image acquisition is where in the input image $I(x, y)$ grayscale image of a palm leaf, it's being captured pixel coordinates xx, yy. Noise reduction is done on the image using a Gaussian filter. A normal map is assigned to the horses. It uses the WHO algorithm in which every horse represents a possible solution. The actual binarization proceeds using Bradley's adaptive threshold method which computes as in the whale optimization algorithm. The fitness of every horse is evaluated in terms of various performance metrics Precision (P), Recall (R), F-measure, Sensitivity (S), Specificity (T), Balanced Classification Rate (BCR), Balanced Error Rate (BER), Geometric Accuracy (G), Precision F-measure (F-measure), and Negative Rate Metric (NRM). Wild Horse Optimization is divided mainly into exploration and exploitation phases. In the exploration phase, the horses move randomly to explore the solution space and update their positions according to Eq. 7.

$$\overline{X_i}(t+1) = \overline{X}_i(t) + \text{rand}\left(\overline{X}_j(t) - \overline{X}_k(t)\right) \tag{7}$$

$\overline{X}_i(t), \overline{X_j}(t)\overline{X}_k(t)$ are positions of horses, and rand is a random number. In the exploitation phase, the best-performing horse attracts others toward its position using the Eq. 8.

$$\overline{X_i}(t+1) = \overline{X}_i(t) + \propto \left(\overline{X}_{\text{best}}(t) - \overline{X}_k(t)\right), \tag{8}$$

where α alpha is the exploitation parameter, and $\overline{X}_{\text{best}}$ represents the best solution so far. To avoid local minima, some horses exhibit random behavior, moving according to Eq. 9.

$$\overline{X_i}(t+1) = \text{rand}\left(\overline{X}_{\min}(t) - \overline{X}_{\max}(t)\right) \tag{9}$$

$\overline{X}_{\min}(t), \overline{X}_{\max}(t)$ are the minimum and maximum parameter values. Figure 8 shows the result obtained by Wild Horse Optimization (WHO) algorithm with Bradley binarization of olaichuvadi (palm leaf manuscript).

In the final stage, however, the method is to be augmented with a great deal of postprocessing: morphological operations, opening, and closing—applied in a highly automated way. When the convergence criterion is tested, either the fitness function stabilizes or the maximum number of iterations is reached to conclude the process; the optimized threshold and window size give way to final binarization. The optimized binarization is then evaluated according to the aforementioned performance metrics. This enables assessment of the quality and accuracy of the binarization process in capturing the existing text while minimizing the incidence of erroneous classification of text versus background. The method proposed is very efficient in retaining palm

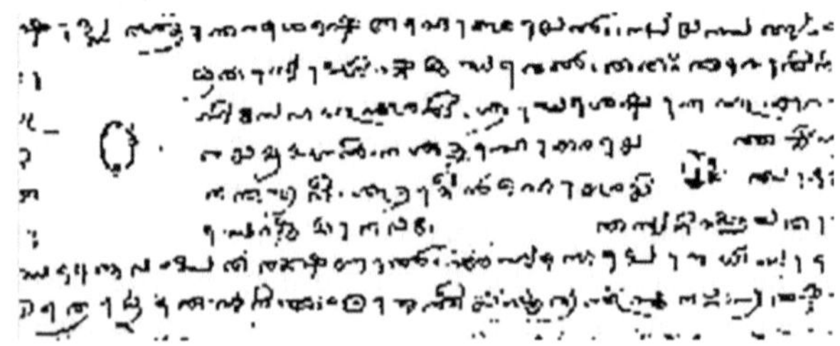

Fig. 8 Wild Horse Optimization (WHO) algorithm with Bradley binarization of olaichuvadi (palm leaf manuscript)

leaf inscriptions and can be extended further to provide binarization for other cases where higher accuracy and precision are required.

5 Whale Optimization Algorithm (WOA) with Bradley Binarization for Image Segmentation and Performance Metrics

The Whale Optimization Algorithm (WOA) with Bradley binarization is proposed for optimal image binarization for palm leaf preservation. The aim is to enhance the clarity of inscriptions on palm leaves by obtaining the optimal binarization threshold and window size through WOA such that a decent distinction between the text and the background is achieved. Binarized images are assessed using Precision, Recall, F-measure, Sensitivity, Specificity, and other applicable techniques to quantify how well binarization works.

Parameter optimization using heuristic algorithms such as WOA and WHO are applied to fine tune the Bradley Binarization process for identifying words in palm leaf manuscripts. Threshold and window size used in binarization are optimized. Figure 9 shows the flowchart of WOA with Bradley binarization. Initially, the olaichuvadi images are $I(x, y)$, where x and y represent the pixel coordinates are acquired, followed by preprocessing, such as noise reduction through Gaussian filtering, described by the Eq. 10.

$$I'(x, y) = \sum_{i=-k}^{k} \sum_{j=-k}^{Nk} I(x + i, y + j)G(i, j), \tag{10}$$

where $G(i, j)$ is the Gaussian kernel and k is the kernel size. This ensures that the text details are preserved while the noise is removed.

The Whale Optimization Algorithm (WOA) starts with a population of whales, each one standing for candidate solutions which consist of binarization threshold T and window size M x N—these seem to be critical parameters to Bradley binarization process. The position of each whale is then measured by a fitness function that comprises binarization metrics such as Precision, Recall, and F-measure. The Bradley binarization for the respective whale is given by the following Eq. 11.

$$T_{\text{Brandley}}(x, y) = (1 - p)\frac{1}{M \times N} \sum_{i=-M/2}^{M/2} \sum_{j=-N/2}^{N/2} I'(x + i, y + j), \tag{11}$$

where p is a constant, typically ranging between 0.15 and 0.2. The objective function is a combination of precision, recall, and $\text{F}_{\text{measure}}$.

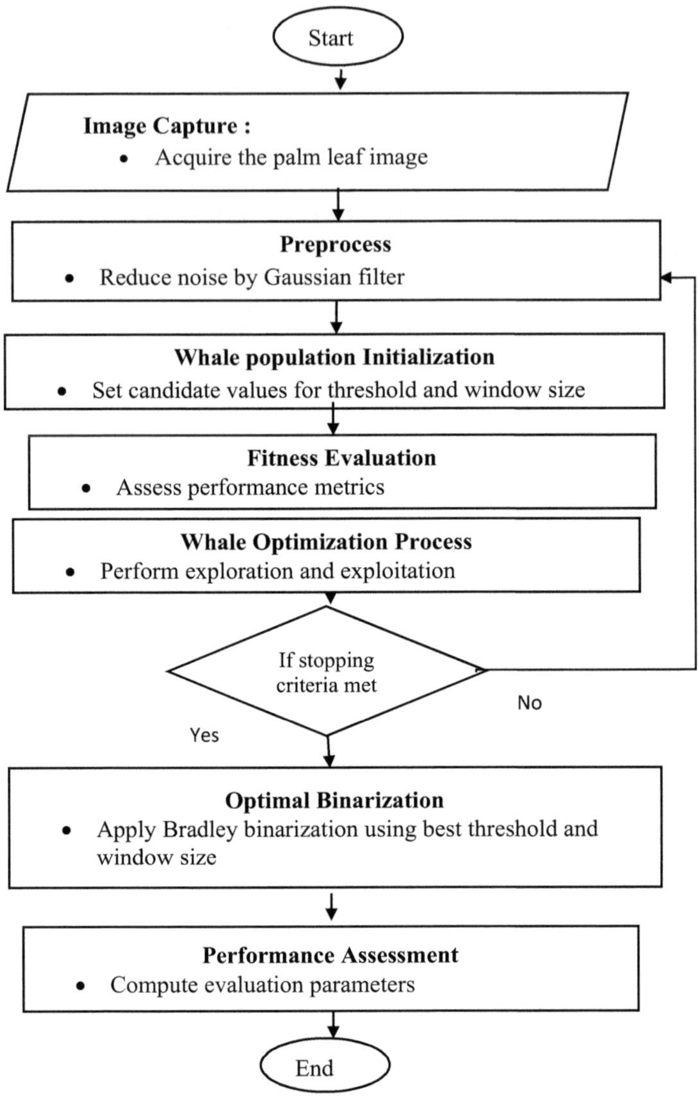

Fig. 9 Flow chart of Whale Optimization Algorithm (WOA) with Bradley binarization

$$\text{Precision } (P)\ P = \frac{\text{TP}}{\text{TP} + \text{FP}} \tag{12}$$

$$\text{Recall } R = \frac{\text{TP}}{\text{TP} + \text{FN}} \tag{13}$$

$$F - \text{Measure} = \frac{2P.R}{P + R} \tag{14}$$

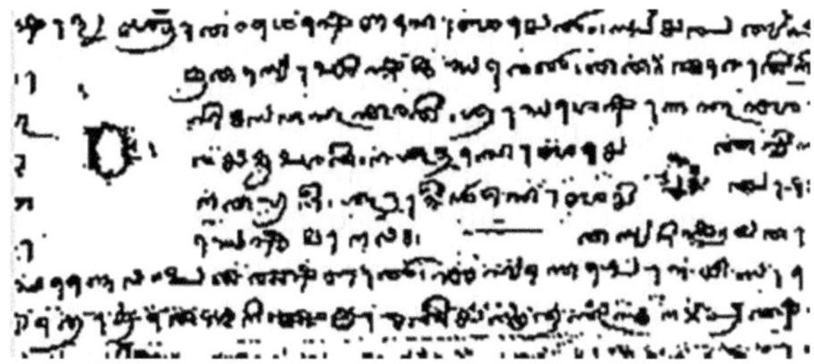

Fig. 10 Whale Optimization Algorithm with Bradley binarization of olaichuvadi (palm leaf manuscript)

The WOA then proceeds with both exploitation and exploration. During exploitation, whales spiral toward the best solution, represented by the best threshold and window size, using the Eq. 15.

$$\overline{X_i}(t+1) = \overline{X}_{\text{best}}(t) + A.\text{D}, \tag{15}$$

where A and D are control parameters, and $\overline{X}_{\text{best}}$ is the current best solution. Exploration is facilitated by updating whale positions based on random distances from the prey, ensuring new potential solutions are explored. This is described by the Eq. 16.

$$\overline{X_i}(t+1) = \overline{X}_{\text{best}}(t) + A.D\overline{X_i}(t+1) - \overline{X_i}(t) = \overline{X_i}(t) + B\sin(2\pi.t), \tag{16}$$

where B controls the spiral shape, and t is the current iteration.

A process keeps running until it meets a stopping criterion. The stopping criterion is either when we find a best threshold and window size or the 100th iteration has been achieved. With optimal parameters in place, Bradley binarization can be applied to the whole manuscript image.

After this step, several postprocessing techniques such as morphological operations (e.g., erosion or dilation) can be applied in order to refine the binarized image. Figure 10 shows the result obtained by algorithm Whale Optimization Algorithm with Bradley binarization of olaichuvadi (palm leaf manuscript).

The effectiveness of binarization is evaluated through performance metrics such as Precision, Recall, F-measure, Sensitivity, Specificity, Geometric Accuracy, F-Measure, Balanced Classification Rate (BCR), and Negative Rate Metric (NRM). Thus, the Whale Optimization Algorithm is adaptive and an intelligent method used to optimize binarization and word identification in palm leaf manuscripts.

6 Results and Discussion

The study compares Whale Optimization Algorithm and Wild Horse Optimization for improving Bradley binarization applied to palm leaf manuscripts. These methods focus on the optimization of the binarization threshold and window sizes for an accurate word identification, and results are analyzed using different performance metrics. This paper discusses the results as per the metrics provided in Table 1. Precision quantifies the proportion of correctly identified words versus all words detected. WOA has a better precision score than WHO, 0.8668 to 0.83556, which shows that WOA slightly reduces false positives. This means WOA distinguishes text from background noise more accurately, usually preserving word structures. Recall checks the power of identification of all actual words within the manuscript.

WOA's 0.76879 is significantly greater than WHO's 0.58507. This shows that WOA is sensitive for finding many actual words in the manuscript, reducing false negatives or missed detections. WHO had a lower recall because many true words in the manuscript went undetected, which was likely because of its relatively ineffective exploration of the search space. The F-measure is a linear combination of precision and recall. WOA has an F-measure of 81.4860% which is higher than WHO's F-measure of 68.82356%, indicating a superior balance between precision and recall for WOA. Thus, WOA works well for binarizing palm leaf manuscripts with fewer overall misclassifications. The sensitivity, the same thing as recall, is higher for WOA (0.76879) than for WHO (0.58507). In terms of specificity, WHO scores better than WOA with a score of 0.98537 compared to 0.97524 in identifying non-word areas. This means WHO is slightly conservative, i.e., it successfully prevents false-positive detections, but fails in the detection of true positives in contrast to WOA. WOA scores 0.87202 for BCR against WHO's 0.78522, hence proving that WOA has a better balance for classification. This metric allows for a confirmation that WOA features overall better performance in relation to word detection and background noise rejection than WHO.

A lower BER means better performance as in the case of WOA: 12.79819% compared to WHO: 21.47757%. This means there are fewer errors that WOA encounters in distinguishing text and backgrounds, and hence WOA gives better performance. WOA has geometric accuracy 0.86589 against WHO's 0.75929. Thus, WOA shows to be more supportive in maintaining the geometric structure of words in the binarization process. This guarantees better shape preservation of words and their forms with WOA. A larger PSNR value leads to better image quality after binarization. WHO scores slightly better with PSNR of 12.23702 against WOA: 12.18085. Although both methods display some decent image quality, WHO has a slight advantage in reducing noise in the binarized image. WOA received higher grades than WHO by a wide margin; recall, F-measure, sensitivity, BCR, BER, geometric accuracy, and NRM are definite outliers and are shown from Fig. 11a–h. This means that WOA boasts a more guarded contention for words in palm leaf manuscripts inasmuch as it manages to balance the tradeoff between successfully detecting words and minimizing the potential for false positives. WHO is slightly better, yet far less effective than WOA on word detection.

WHO works better for specificity and most likely PSNR does not favor it in word detection compared to WOA, as WHO emphasizes crime or as they would call it avoidance of false-positive instances, thus preventing WHO from sufficiently gnawing through the solution space to return decent recall and F-measure. Both optimization techniques show promise in bettering the binarization process where

Table 1 Various Performance metrics of olaichuvadi using thresholding techniques

Performance Metric	Wolf Binarization	Niblack Thresholding	Sauvola Thresholding	Bernsen Thresholding	Nick Thresholding	Bradley Thresholding	Whale Optimization Algorithm	Wild Horse Optimization
Precision	0.24706	0.48878	0.34981	0.58275	0.70502	0.79637	0.866	0.83556
Recall	0.02877	0.47325	0.02922	0.33224	0.34406	0.57130	0.76879	0.5850
F-measure (%)	5.15337	48.08934	5.3927	42.3201	46.24462	66.53200	81.4860	68.8235
Sensitivity	0.02877	0.47325	0.029	0.33224	0.3440	0.57130	0.76879	0.58507
Specificity	0.99603	0.69657	0.9963	0.868	0.98253	0.97222	0.97524	0.9853
BCR	0.51240	0.58491	0.512	0.600	0.663	0.77176	0.87202	0.78522
BER(%)	48.7603	41.5086	48.719	39.95781	33.67024	22.82356	12.7981	21.47757
Geometric Accuracy	0.169	0.57416	0.170	0.537	0.58142	0.74528	0.86589	0.75929
NRM	0.487	0.41509	0.487	0.39958	0.33670	0.22824	0.12798	0.21478
PSNR	13.38145	1.10836	11.945	4.91818	10.62667	10.3702	12.18085	12.2370
DRD	0.959	12.19766	1.327	9.575	2.54867	2.72523	2.2544	2.36686
Best Score	0.069	0.195	0.077	0.169	0.086	0.087	0.07446	0.07334

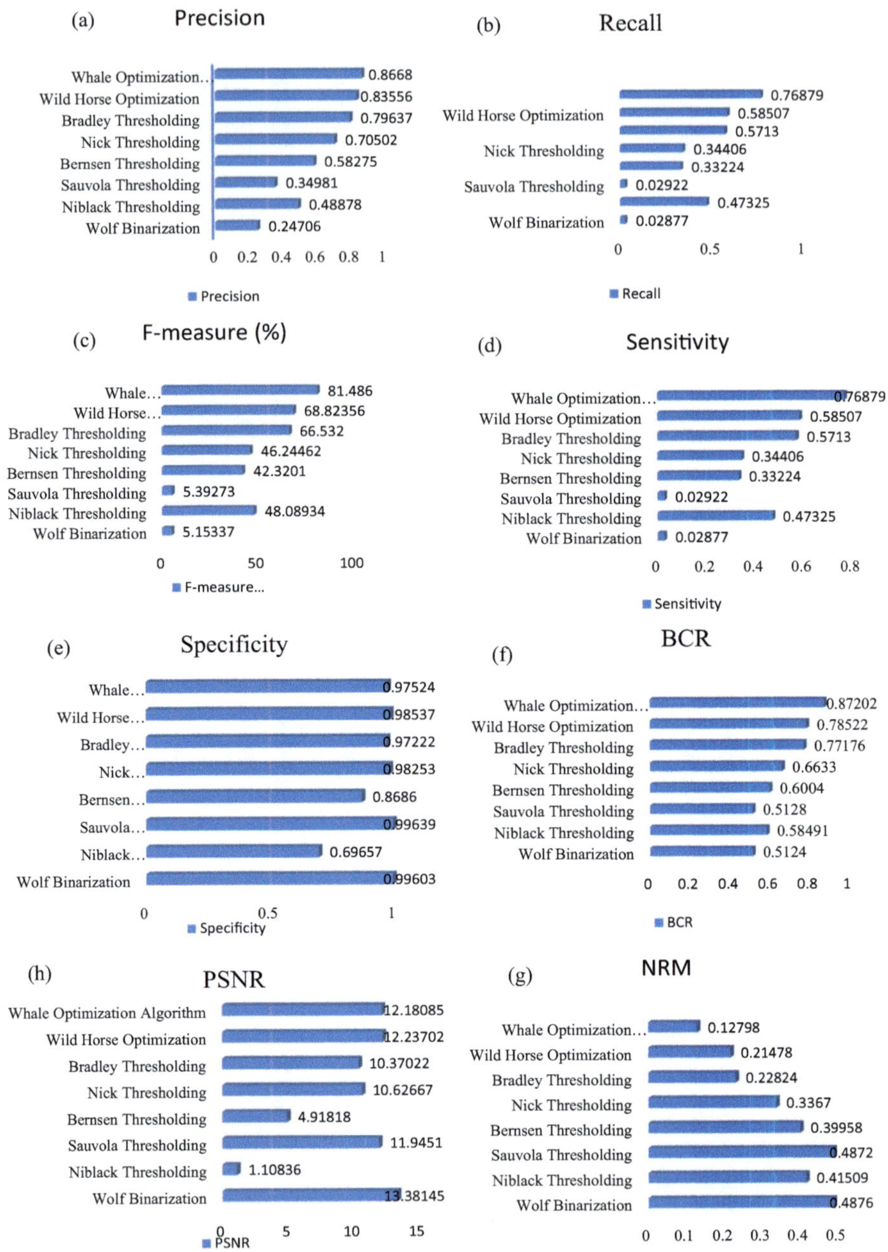

Fig. 11 **(a–h)**: Performance metrics Precision, Recall. F-Measure, Sensitivity, Specificity, BCR, PSNR, NRM for olaichuvadi using thresholding techniques

WOA demonstrates a better performance in terms of overall performance metrics. The determined results provide that WOA is even more effective in the area of palm leaf manuscripts in achieving better word recognition and maintaining. The olaichuvadi system ensures that ancient knowledge is not just digitized but made intelligently accessible, particularly to communities that have traditionally preserved and practiced these texts. Artificial Intelligence provides cultural resilience, supporting equitable development and social cohesion in line with global SDG principles.

7 Impact on Sustainable Development Goals

Restoration of olaichuvadi is required for sustainable development as it holds cultural and historical information. With ever-growing artificial intelligence is used to harness the wisdom in ancient script and bridges the gap to the modern technology offering scalable and inclusive solutions to safeguard intangible heritage. In alignment with the United Nations Sustainable Development Goals (SDGs), this system facilitates equitable access to indigenous knowledge, fosters innovation, and strengthens educational and research ecosystems. The impact of olaichuvadi across selected SDG targets is summarized in Table 2.

The system addresses SDG 11.4 by preserving olaichuvadi in term treasuring the knowledge wealth in digital means. Real-time access with inclusive information in local language enhances SDG 4.7 by clubbing cultural heritage with education for sustainable development. The system's use of edge AI and low-cost deployment fosters SDG 9.5, encouraging innovation in underrepresented domains. Furthermore, olaichuvadi strengthens SDG 17.6 and 17.8 by encouraging technological partnerships and making heritage technologies accessible to rural and marginalized communities. This holistic impact reinforces the importance of AI not only as a tool for efficiency but also as a catalyst for equitable, sustainable knowledge preservation.

Table 2 Contribution of olaichuvadi system to SDGs

SDG Target	Impact of AI-Driven Olaichuvadi System
SDG 11.4	Protects and promotes cultural heritage through AI-based restoration and preservation
SDG 4.7	Supports education for sustainable development by making indigenous texts accessible
SDG 9.5	Fosters innovation and digital research infrastructure in heritage domains
SDG 17.6 & 17.8	Promotes cross-disciplinary partnerships and inclusive technology access

8 Conclusion

Two metaheuristic algorithms—WOA and WHO—were compared with respect to their performance in improving binarization techniques in the digitization and preservation of ancient manuscripts, in this case, the olaichuvadi of palm leaf manuscripts. The importance of the binarization process is that it improves the legibility of degraded manuscripts, where problems arise from uneven illumination, noise, and text degeneration. Conventional global thresholding techniques often perform poorly; hence adaptive binarization methods with optimization algorithms emerge as a better choice. It was witnessed that the WOA consistently outperforms WHO over many performance measures considered for this assessment, notably in terms of Precision (0.8668), Recall (0.76879), and F-measure (81.4860%), which ensure WOA does great in distinguishing between text and background noise while preserving word structures. Furthermore, the geometric accuracy of WOA (0.86589) proved better than WHO and its more than convincing balanced classification rate (BCR) of 0.87202 established its competency in separating text regions while maintaining the geometrical structure of words. Lower bit error rate (12.79819%) for WOA also ensured that it produced fewer errors while segregating text from non-text regions. Though WHO had minimal value in specificity at 0.98537 with a peak signal-to-noise ratio (PSNR) of 12.23702, it still had far less call in word detection, with lower Recall (0.58507) and lower F-measure (68.82356%). Thus, it can be interpreted that WHO focused mainly on avoiding false positives, while it might have lived out effective identification of actual words. In brief, although both strategies have been found effective for improving binarization processes, WOA emerges as the better and fitter algorithm. Its performance on major performance indicators recommends it as an adequate candidate for improving the readability and restoration of ancient manuscripts. This study thereby demonstrates the possible efficacy of WOA as an efficient technique in cultural heritage conservation via digitization. Olaichuvadi showcases the synergy between AI and cultural sustainability, offering a scalable model for heritage digitization. Future enhancements include multilingual semantic search, cloud-based manuscript repositories, and integration with AR/VR platforms for immersive heritage learning.

References

1. Alexander, T.J., Kumar, S.S.: A novel binarization technique based on Whale Optimization Algorithm for better restoration of palm leaf manuscript. *J. Ambient Intell. Humaniz. Comput.* **11**, 1–8 (2020)
2. Poornima Devi, M., Sornam, M.: Feature engineering for Tal-Patra manuscript text using natural language processing techniques. In: Mandal, J.K., Chaki, R., Banerjee, S. (eds.) *Proc. Third Int. Conf. Sustainable Computing: SUSCOM 2021*, pp. 279–287. Springer, Singapore (2022)
3. Wiland, J., Brown, R., Fuller, L., Havelock, L., Johnson, J., Kenn, D., Snowdon, J.: A literature review of palm leaf manuscript conservation—Part 1: A historic overview, leaf preparation,

materials and media, palm leaf manuscripts at the British Library and the common types of damage. *J. Inst. Conserv.* **45**(3), 236–259 (2022)
4. Nishanthi, M.: Traditional conservation methods used to preserve Indian palm leaf manuscripts: A literature review. (2023)
5. Chu, S., Lin, L., Tian, X.: Evaluation of the deterioration state of historical palm leaf manuscripts from Burma. *Forests* **14**(9), 1775 (2023)
6. Lertratanakehakarn, P.: Digital preservation of palm-leaf manuscripts in Thailand. In: Zaphiris, P., Ioannou, A. (eds.) *The Emergence of Digital Libraries – ICADL 2014*, pp. 12–21. Springer, Cham (2014)
7. Michalak, H., Okarma, K.: Robust combined binarization method of non-uniformly illuminated document images for alphanumerical character recognition. *Sensors* **20**(10), 2914 (2020)
8. Michalak, H., Krupiński, R., Lech, P., Okarma, K.: Preprocessing of document images based on the GGD and GMM for binarization of degraded ancient papyri images. In: Choraś, M., Choraś, R.S. (eds.) *Computer Recognition Systems*, pp. 116–124. Springer, Cham (2021)
9. Bloechle, J.L., Hennebert, J., Gisler, C.: ZigZag: A robust adaptive approach to non-uniformly illuminated document image binarization. In: *Proc. ACM Symp. Document Engineering*, pp. 1–10 (2024)
10. Lins, R.D., Bernardino, R.B., Barboza, R.D.S., Simske, S.J.: Binarization of photographed documents: Image quality, processing time and size assessment. In: *Proc. 22nd ACM Symp. Document Engineering*, pp. 1–10 (2022)
11. Amiriebrahimabadi, M., Mansouri, N.: A comprehensive survey of feature selection techniques based on whale optimization algorithm. *Multimed. Tools Appl.* **83**(16), 47775–47846 (2024)
12. Lin, Z.: Optimizing Kernel Extreme Learning Machine based on an enhanced adaptive Whale Optimization Algorithm for classification task. *PLoS One* **20**(1), e0309741 (2025)
13. Zheng, R., Hussien, A.G., Jia, H.M., Abualigah, L., Wang, S., Wu, D.: An improved wild horse optimizer for solving optimization problems. *Mathematics* **10**(8), 1311 (2022)
14. Ju, R.Y., Lin, Y.S., Jin, Y., Chen, C.C., Chien, C.T., Chiang, J.S.: Three-stage binarization of color document images based on discrete wavelet transform and generative adversarial networks. *Knowl.-Based Syst.* **304**, 112542 (2024)
15. Pravesjit, S., Seng, V.: Segmentation of background and foreground for ancient Lanna archaic from palm leaf manuscripts using deep learning. In: *2021 Joint Int. Conf. Digital Arts, Media and Technology (ICDAMT) and ECTI Northern Section Conference*, pp. 220–224. IEEE (2021)
16. Subramani, K., Subramaniam, M.: Creation of original Tamil character dataset through segregation of ancient palm leaf manuscripts in medicine. *Expert Syst.* **38**(1), e12538 (2021)
17. Paulus, E., Burie, J.C., Verbeek, F.J.: Text line extraction strategy for palm leaf manuscripts. *Pattern Recognit. Lett.* **174**, 10–16 (2023)
18. Sudarsan, D., Sankar, D.: A novel complete denoising solution for old Malayalam palm leaf manuscripts. *Pattern Recognit. Image Anal.* **32**(1), 187–204 (2022)
19. Haritha, J., Balamurugan, V.T., Vairavel, K.S., Ikram, N., Janani, M., Indrajith, K.: CNN based character recognition and classification in Tamil palm leaf manuscripts. In: *2022 Int. Conf. Communication, Computing and Internet of Things (IC3IoT)*, pp. 1–6. IEEE (2022)
20. Thuon, N., Du, J., Zhang, J.: Improving isolated glyph classification task for palm leaf manuscripts. In: *Int. Conf. Frontiers in Handwriting Recognition*, pp. 65–79. Springer, Cham (2022)

Perspectives on AI-Enabled Education for Sustainable Development

Apramita Chand, Satyam Sarthi, Aanchal, and Gokulvinayak C. R.

Abstract The multi-level percolation of artificial intelligence (AI) into the education sector has revolutionized pedagogical processes worldwide. This chapter explores the transformative impacts of AI-enabled education that range from daily curricular transactions, creation of immersive learning environments to intelligent tutoring systems, and predictive learning analytics from a sustainability perspective. Case studies on AI implementation in education (AIEd) in diverse didactic settings have been analyzed across subject domains to examine the long-term potential of AI-integrated teaching-learning processes. It is expected that such preventive and adaptive pedagogical interventions might encourage metacognitive scaffolding, stimulated and interactive learning through tailored systems, and responsible data mining. In this chapter, we also discuss the effects of leveraging AI for promoting inclusivity and accessibility that can cut across linguistic, geographical, and ethnic barriers, thus providing a bias-resilient educational framework for achieving SDG goals. As lines between the real and the artificial begin to blur, it is also important to have a futuristic glimpse into the ethical challenges posed by unregulated usage of AI-ML techniques in pedagogical processes and how these can be attenuated by periodic assessments by stakeholders. It is expected that these discussions will also guide perspectives regarding mitigating issues of costs, scalability, and standardized guidelines in the context for sustaining AI-enabled education in the long run.

1 Introduction

The world is slowly gearing up for the dawn of Society 5.0 where the virtual and the physical spaces will converge to create a higher-order thinking society that is accepting of the co-existence of digital and human entities working together for a better future. In this context, the thread connecting Society 5.0 and sustainability

A. Chand (✉) · S. Sarthi · Aanchal · Gokulvinayak C. R.
Department of Education, Indian Institute of Technology Kharagpur, Kharagpur, West Bengal, India
e-mail: apramita.chand@edu.iitkgp.ac.in

A. Khanna et al. (eds.), *Artificial Intelligence for Sustainable Development Goals*,
https://doi.org/10.1007/978-3-032-28217-0_11

goals wraps closely around the foundations of Industry 4.0 and Education 4.0 in which artificial intelligence has emerged as a leading force, as illustrated in Fig. 1.

Education for Sustainable Development (ESD) was a concept brewed in the Earth Summit of 1992 in Rio de Janeiro that finally found acceptance in the World Summit of Sustainable Development in 2002. The decade of 2005–2014 was recognized as the United Nations Decade of Education for Sustainable Development (UNDESD). Ever since, ESD has been an integral part of planning for national policy frameworks and has been included in many schemes for the achievement of Millenium Development Goals (MDGs). Considering the 2030 Agenda for Sustainable Development, a human-centered approach is the focus of all initiatives including ESD that calls for transformation of the education sector to cater to lifelong learning, universal access, innovations, and collaborations across boundaries with cultural appropriation while keeping our eyes fixed on global targets.

The all-encompassing AI revolution has swept across all arenas of education, making it an epoch-making era in ESD [1]. The emergence of Industry 4.0 that has created a demand for digitally skilled workforce has its answer in AI-enabled education (AIEd). Unsurprisingly, there have been huge investments of late by UK, USA, and Indian government in revamping AI in education by setting up centers of excellence, building AI platforms, and LLMs, as well as providing impetus to research in artificial intelligence.

Here, we explore the potential of AI-enabled education across different pedagogical avenues and aspects dealing with inclusivity and foreseeable challenges along with futuristic recommendations for a sustainable vision for AIEd [2–8].

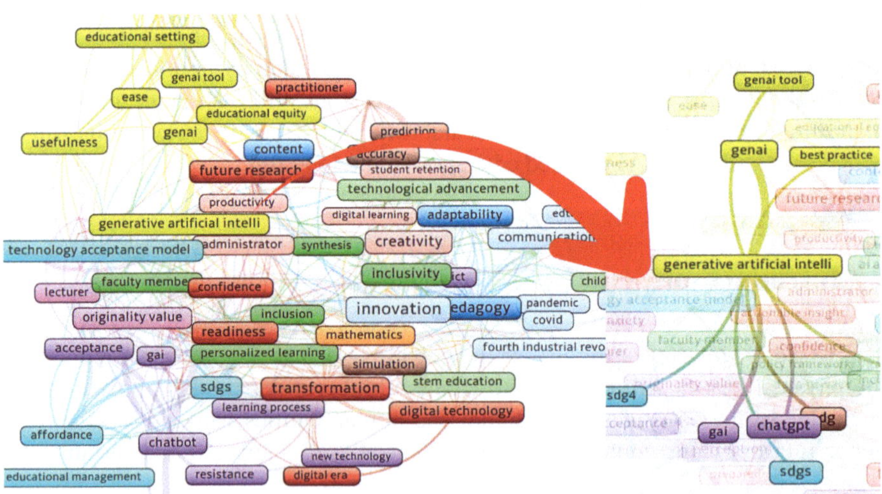

Fig. 1 Thematic map created (using VOSviewer version 1.6.20) from the bibliographic matter of a sample of 340 papers from Dimensions database (https://app.dimensions.ai/). The cluster of generative AI is zoomed into and highlighted to show strong connections with SDGs and best practices

2 Multidimensionality of AIEd

An important aspect of sustainable education is that it aims at transformative outlooks at the three domains of content, process, and context in pursuit of quality education suited to the present and future needs of learners across the world. Sustainability in education is a dynamic concept where technologies, practices, and methodologies are constantly evolving to develop competencies and broaden educational outcomes holistically. The multi-dimensional nature of artificial intelligence promises to cater to diverse learning styles, promote higher-order thinking skills, and deliver positive educational outcomes. If we analyze themes relating to AI in sustaining education, we come across keywords driving at innovation, adaptation, and security, as shown in Fig. 1. We have considered only particular bibliometric data that had keywords "artificial intelligence," "education," and "sustainable" in the title and abstract. Only journal articles and proceedings from 2013–2025 pertaining to "Education" Research Area and inclusive of Sustainable Development Goal 4 (Quality Education) were filtered from the database. Generative AI applications as shown are inextricably linked to SDGs when the linkages of term co-occurrences in literature are established in form of clusters.

2.1 Curating Experiences in the Classroom and Beyond

A wave of generative AI applications has been transforming pedagogical experiences for students and teachers. From learning outcomes designed as per Blooms' taxonomy [9, 10] covering the arenas of Knowledge, Comprehension, Application, Analysis, Synthesis, and Evaluation, to achieving 5E goals in curricular transactions [11, 12], AI has served as a handy tool for upgradation of teaching methods, diversification of learning types, and the evolution of smart classrooms, all while promoting creative engagement. Here, we examine some notable examples of how traditional classrooms have evolved to make way for multi-faceted advanced learning environments peppered with artificial intelligence. A word cloud in Fig. 2 showcases some of the versatile and widely used tools and platforms in AIEd.

2.1.1 Lesson Planning, Creation, and Delivery

AI tools and platforms are helping teachers prepare well-organized lesson plans for conducting smoother learning sessions by systematically arranging content elements in pedagogical epistemic pathways, integrating audio-visual stimuli, designing activities, anticipating outcomes, challenges and offering solutions. For example, To-Teach AI is a tool that has been explored by EFL (English as a Foreign Language) teachers to clearly define instructional objectives, tailor worksheets, and generate rubrics for efficient assessment [13]. An intelligent AI software tool, CLEVER [14],

Fig. 2 WordCloud of Versatile AI Tools/Platforms available in the public domain used for various curricular transactions

has been developed for preparing didactically rewarding lesson plans by customizing arrangement of content according to competencies and is being tested in pilot mode in different German schools. Similarly, CollectiveTeach [15] is a platform that has been designed as a document retriever and sequencer of relevant information taking input of simple keywords from the user and generating accurate lesson plans using a concoction of mixture models, k-means clustering, and spectral clustering techniques. Large language models like ChatGPT have been extensively used as a one-stop solution for creating customized lesson plans swiftly, brainstorming ideas for activities and group discussions, devising simple quizzes, and comprehensively outline instructional objectives, saving teachers time and effort [16–22]. The quality of input received from the educators has been shown to dictate the accuracy level of the solutions provided, in which case high specificity can be achieved by providing standardized curriculum texts along with clear instructions as input [23].

The UDL framework or Universal design for learning has accepted guidelines for correct representation, engagement, and expression in educational content and AI-generated lessons with immersive elements like graphics, videos, quizzes align with UDL wireframe [24].

AI-powered data analytics can now tune delivery of educational content according to student preferences like type of presentation, speed of learning, preferences, and even performance levels. This is known as adaptive learning that follows a looping architecture of collecting feedback and responses and utilizing it to tailor different kinds of content in specific sequences depending on the learner's skills, goals, interactions with the chosen platform, strengths, and weaknesses. These systems also learn to curate activities to promote learner engagement and offer individualized feedback [25]. By gauging emotions of a student crowd based on parameters like facial expressions, body language, and pupil dilation, teachers in AI-enabled smart classrooms can be guided to modify their own presentation styles in real time [26].

2.1.2 Student Engagement

Sustaining student engagement has become a real challenge with students struggling to remain invested in the learning process owing to hectic schedules, information overload, and diversified learning styles. As educators move away from traditional lecture-centric models to immersive learning environments, personalization and interactivity are two crucial avenues that are being actively addressed by AIEd designers (Fig. 3).

AI-generated pedagogical agents (PA) based on Generative Artificial Intelligence (GAI) that create relatable humanized characters and settings for augmented reality (AR) environments have been shown to improve affective domains in learners establishing emotional connections that provide better immersion in the lessons. The typical setup of such GAI-assisted AR platforms includes attractive animations, 3D models, simulations, tutorials, and feedback videos. A case study with 60 students studying the concepts of solar system was presented with DongDong—a GAI-based PA, assisting students through various modalities of the lesson and positive correlations were drawn with respect to student interest and understanding [27]. Gamified immersive learning environments (see Fig. 4a) are possible through AI-assisted AR/VR technologies as encapsulated in Table 1 that highlights some of the prominent examples in this field.

AI teaching aids have also made use of virtual simulation platforms to gauge student-platform interaction behavior and promote participatory learning in groups or as individuals [28]. AI-based AR/VR immersive experiences created by Nearpod VR, Google Expeditions, and Merge Cube have been beneficial for students to visualize historical places, body organs, biological processes. Adaptive learning provided by Knewton and Dreambox offers customized learning pathways based on student interactivity patterns and test scores, thus exploring more pathways for capturing student attention in the right way.

2.1.3 Assessment and Evaluation

AI-enabled automated grading of assignments, particularly for objective type questions, as well as syntax and grammar checking of reports and project essays can reduce teacher workload [37]. Plagiarism detection software Turnitin now has AI-enabled features based on natural language processing to spot discrepancies in written materials and copied sentences. Grammarly [38] has advanced AI features embedded for correcting and improving writing styles.

AI-powered platforms like Quizziz (snapshot shown in Fig. 4b), Quizlet, Padlet, Socrative, and Formative have modules for preparation of revision sets, flashcards, and practice sets including different kinds of questions ranging from multiple choice,

Table 1 Examples of AI-assisted AR/VR applications along with their impacts on teaching–learning processes

Application/Platform	AI technology used	Educational impact
"You, Calligrapher" VR Learning [29]	Natural Language Processing and Image Processing, Virtual Teacher Assistants	Effectiveness Remote level art engagement Immersive virtual learning
AR Origami Training System [30]	BGM Deep learning (YOLOv8) Segment Anything Model (SAM) by Meta AI, along with Track Anything algorithm used.	Cognitive awakening in university students Development of informal skills
AR Precision Manufacturing Tutor [31]	Object recognition, Machine learning for user modeling	Reduced assembly errors by 31–84%; engineering students showed increased competency step-by-step guidance with adaptive feedback.
"AnimalCircle" AR App [32]	Deep Learning Image recognition as AR content trigger	Primary school science education Mobile app for school students Overlay of AR images and facts
AIEduAR [33]	Gj Supervised learning techniques to train data, data classification and Computer Vision (QR Code Recognition	Problem solving using machine learning
LearningverseVR [34]	Large Language Model Stable Diffusion (Text-to-image generation)	Collaborative learning Virtual Gamified Environment Personalized interactions, Conversation and emotion.
Smart Campus Learning App [35]	Adaptive Learning, AI + Internet of Things (AIoT)	Context-aware AR platform Situation-based learning University campus as sample setting Interactive media-rich interface
Interactive platform for Remote Practical Works (RPWs) [36]	Personalized learning through machine learning algorithms AR/VR experiences	Participatory distance learning, Virtual lab management for students

match the words, true/false, fill in the blanks to test learner progress. The inputs can vary from textual inputs to standardized document PDFs and video URLs for generation of quizzes [39]. Many of these platforms offer gamified experiences for promoting student interest and critical thinking.

Large Language Models [40, 41] like ChatGPT, Google Bard, Gemini AI, Microsoft Bing have been efficient in creation of grade-specific testing rubrics, grading guidelines, appraisals, recommendations, individualized reports for learners, and personalized letters to parents.

The trustworthiness of peer assessment can be raised by assistance from AI-based models like RiPPLE that can provide guidelines, accuracy levels, and resource text matching, along with modulating quality of feedback [42].

2.2 Smarter Educational Administration

Administrative tasks in education include multi-level decision-making, execution of academic and non-academic tasks relating to hiring, time table scheduling, student data handling, and management along with management of examinations and admissions. Human-centered AI-based approaches can help streamline these processes and increase operational efficiency while reducing the margin of error [43–45].

In North Carolina, student records were assessed in the Department of Public Instruction, based on 74 metrics of behavior, socio-economic status, performance and demographics and machine learning techniques were applied [46] to present risk analysis of potential drop-outs. Similarly, data of 72,598 Danish pupils pursuing secondary education, obtained from online sources, Lectio study administration system used in Danish high schools as well as government statistics, was processed through a random forest classifier to give 93.47% accuracy in predicting dropout rates. These could be sustainable solutions for developing a warning mechanism for the school administrators to target identified groups of students for extra coaching/motivational programs. Learning analytics accessed by student center in University of South Florida has been useful in delivering targeted interventions to students with learning difficulties.

In a research study by Kawamura et al. [47], data collected through multimodal means like facial data, seat pressure, pace of heartbeats (shown in Fig. 4c) was analyzed statistically and algorithms like Random Forest, CatBoost, and multiclass Support Vector Machine were used to train machine learning models to automatically detect levels of wakefulness and drowsiness in learners interacting with e-learning platforms. This could be used in alert systems as well as feedback loops.

Educational leadership can be motivated by a symbiotic human–AI engagement for evidence-based decision-making from educational data mining [48, 49]. These real-time data recommendations can influence decisions regarding budget allocations, manpower recruitment, student academic programs, achievement rates, disciplinary action, and replenishing of stock inventories.

Learning Management Systems (LMS) are now equipped with AI-powered features that enhances student activity, analytics, management of data, along with collaborative features. In short, AI-driven automation can streamline the organization and logistics of learning curves within an LMS. A study [50] conducted across Saudi Arabian universities with 1991 participants used structural equation modeling to derive positive correlation between automation of learning management systems using artificial intelligence and readiness of students toward e-learning which points toward the potential of AI for designing intuitive interfaces for educational administration. Techniques of AI have been integrated within a React-based web application to develop a school management system [51] with suggestion of teacher strategies, learning workflows, automatic tracking of attendance, class scheduling, performance indicators, and automation of management tasks with real-time processing of data. It provides a sustainable framework for the future with propositions for added adaptability, interactivity, and scalability.

Forecasting by pattern detection, automation of data analysis, classification, screening, and back-end office operations can reduce human effort and error in educational administration which can be explored by advanced synergy between AI and Robot Process Automation which is set to revolutionize several sectors, in line with the vision of Industry 4.0 [52].

2.3 *Use of AI Chatbots*

It has been demonstrated [53] that students can interact with AI chatbots in multiple ways [54–58] to embark on self-regulated learning pathways. They can engage in chat for brainstorming ideas, preparation of step-by-step flowcharts, and doubt clarification. Students can also frame their rubrics for self-assessment and ask the chatbot to evaluate their work. They can further engage in discussion about the assessment procedures. This approach enhances learning through cues, a key component of the Judgement of Learning (JOL) framework, which is often encountered in educational psychology studies. This type of continuous learning where the chatbot asks the learner to think from different perspectives or reflect deeper can help learners be conscious of their learning journeys thus aiding metacognitive abilities.

In a meta-analysis study of the effect of chatbots in education, it was observed that, irrespective of moderating variables, over 2201 participants showed medium-to-high improvements in learning outcomes across several parameters [59]. Chatbot-assisted feedback has been proven to be coherent, detailed, and naturally personalized, thus achieving nearly the same result as a human instructor [60].

When we examine whether AI-powered chatbots would be sustainable, we can consider a study by Al-Sharafi [61] where factors like perceived usefulness, satisfaction, knowledge acquisition, and application were weighed against responses received from 448 university students and a positive correlation between these factors and continued use of chatbots by students was found through a combination of structural equation modeling and artificial neural networks.

Not surprisingly many internationally acclaimed universities have launched their own chatbots [62] like Arizona State University (chatbot named "Sunny"), Washington University ("Martha"), Staffordshire University ("Beacon"), Georgia State University("Pounce"), Georgia Institute of Technology ("Jill Watson"), Complutense University of Madrid("Lola"), and Polytechnic University of Valencia ("Pau") that fulfill thousands of multifarious requests everyday ranging from students queries on admission, registration, course selection, subject content, and doubt clarification to functioning as teaching assistants.

2.4 Intelligent Tutoring Systems

Intelligent Tutoring Systems are intuitive models built to function as personalized tutors for students by applying techniques of artificial intelligence for customized learning pathways, content presentation, adaptive techniques for assessment, and gauging of performance along with one-on-one recommendations for improvement.

A relevant ITS must possess knowledge of the domain, knowledge of the learner, and knowledge of teaching strategies. It has been seen that a student learns from an ITS primarily by solving problems and interacting with the system that leads to the system assessing what the students knows and concurrently considering what the student needs to know further in the curriculum. Finally, the ITS decides which curriculum element ought to be instructed next and how it shall be presented. From all of these considerations, the system selects or generates a problem, and then either works out a solution to the problem or retrieves a prepared solution. These processes are integrated in the basic architecture of any ITS [63] as shown in Fig. 3.

Using Natural language processing, ITS like IBM's Watson Tutor can extend students' understanding of the subject by clarifying doubts in math and science. Similarly, NLP-based "on-the fly" adaptive feedback generation has been found to be useful in case of The Basic Electricity and Electronics Tutorial Learning Environment (BEETLE II), highlighting importance of feedback specific to the context riding on correct interpretation of student responses. Collaborative efforts between Google and Khan Academy also offer intelligent tutoring support and personalized recommendations based on performance of the learners. Further, a student-centered end-to-end solution in form of a web-based ITS called ZOSMAT [64] has been developed (shown in Fig. 4d) that is not only inclusive but also comprehensive as it includes adaptive content delivery (in form of attractive interactive content like animations), conduct of examinations, performance evaluation, individual reports, and analytics.

A Java-based ITS and virtual simulation environment, Virtual Physics System (ViPS) [67] has been evaluated by coached problem-solving approach with 200 students from higher education institutions to test and remedy their knowledge of pulley systems. Logic tutor systems like DeepThought [68] have been envisaged using Bayesian networks to improve student proficiency in problem-solving statements and logical thinking for computational applications. Collaborative learning approaches have been enhanced using a multi-agent tutoring system called Í-MINDS [69] using the Jigsaw model of learning along with VALCAM algorithm to foster group learning exercises and monitoring.

Students performance and reliance on teacher help in after-school settings was seen to be improved in 253 participants after continuous exposure to the Assessment

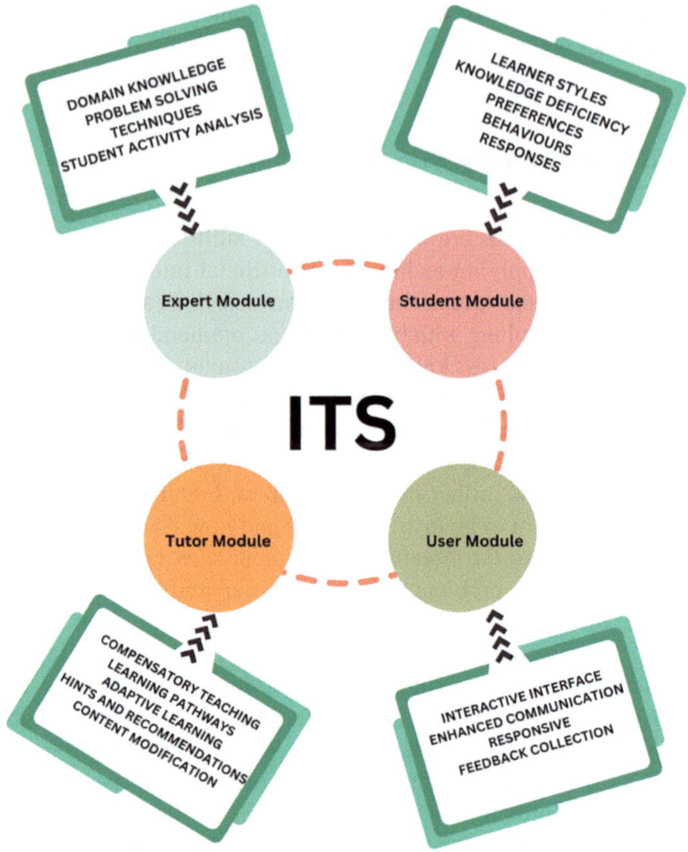

Fig. 3 Basic architectural composition of an ITS with four different modules encompassing different verticals [63]

and Learning in Knowledge Spaces (ALEKS) ITS system in which mathematics was focused as the learning competency to be developed [70].

An AI-based tool aligned with principles of UN 2030 Agenda SDG with integration of OpenAI API chatbot in the interface was designed by Dominguez and group [71] to provide personalized instruction based on learning styles of engineering students over a 16-week-long semester utilizing long short-term memory (LSTM) neural network. This was done to include preferences of learners categorized within the four-fold dimensionalities (visual/verbal, sensory/intuitive. Sequential/ global, and active/reflective) of Felder–Silverman Learning Style Model (FSLSM).

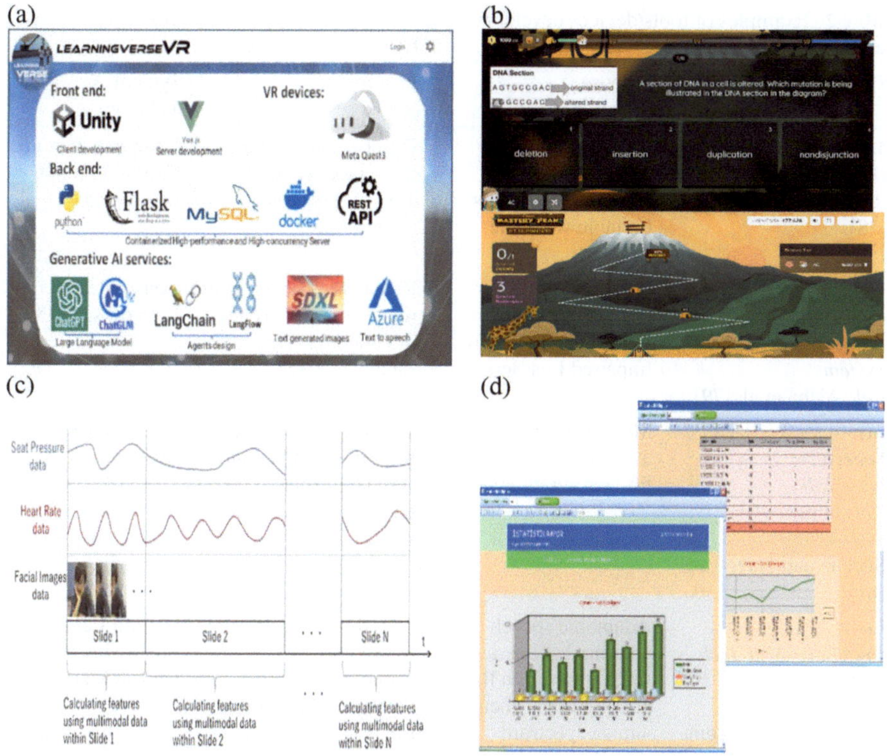

Fig. 4 **a** Representation of Learnerverse VR setup. Reproduced from ref. [34] under Creative Commons License (https://creativecommons.org/licenses/by/4.0/). **b** Snapshot of Mastery Peak Quiz and Dashboard of Quizziz AI assessment platform, **c** Detection of heartbeats, pressure, facial data for monitoring attentive behavior of learners. Reproduced from ref. [65] under terms of Creative Commons Attribution 4.0 International (https://creativecommons.org/licenses/by/4.0/), **d** ITS ZOSMAT showing learner reports interface. Reproduced with permission from ref. [66]. Copyright 2009 Elsevier

3 Inclusivity in AI-Enabled Education

Assistive technologies are a boon for people with disabilities as these allow their functional abilities to be enhanced while allowing them to pursue learning independently [72]. AI-driven assistive technologies have made use of data analytics and sophisticated algorithms to make inclusion of specially challenged learners seamless by ensuring better access and flexibility of learning [73]. For preparing accessible solutions, AI frameworks—whether deployed in websites or apps, must be in tandem with guidelines of Universal Design for Learning (UDL) [74] and the Web Content Accessibility Guidelines (WCAG) [75]. Examples of inclusive AI tools have been highlighted in Table 2.

Table 2 Examples of tools/devices developed using artificial intelligence to assist learners facing various kinds of disabilities

Tool/App	Purpose	Features	Type of AI technology
Microsoft's Seeing AI *OrcamMyEye* Ref: Granquist et al. [78]	App for vision accessibility for classroom activities Portable wearable device	Describes text, auditory cues, face recognition Text to speech recognition	Computer Vision, Natural Language Processing
Virtual Sign Language Recognition and Interpretation (VSLRI) System Ref: Aicha et al. [79]	Sign Language Recognition for Hearing and Speech Impaired Learners	Real time processing of image and video cues in American Sign Language (ASL)	Convoluted Neural Networks (CNN)
Ava *Speechify* *Let's Talk* Ref: https://www.ava.me/ [80] Seebun and Nagowah [81]	Mobile app-based assistive technologies for hearing impaired	Transcribes text live, has lip reading features, sign language recognition	Visual Speech Recognition
Logan ProxTalkerMid-Tech AAC device Ref: https://logantech.com/products/proxtalker [82] Pak et al. [83]	Adaptable speech communication device for learners with the hearing impairment, autism spectrum disorder	Sound tagging, Picture assistance, Radiofrequency identification (RFID) system	Natural Language Processing with Pattern Matching
DysLexML *ALEXZA* Ref: Asvestopolou et al. [84] Rajapakshi et al. [85]	Screening tool for dyslexia Digital Support Tool for dyslexic learners	99% dyslexia diagnosis from eye tracking Picture–Text recognition, read-aloud features	Support Vector Machine (SVM) Feedforward neural networks, AI Markup Language
Taylor Ref: Lister et al. [86]	AI assistant for managing university administrative tasks for disabled students	Form-filling, chat support, accommodation procedures	Voice recognition and synthesis

Smart AI assistants in the form of mobile applications have been developed using Cloud Vision API [76] to function as object recognition systems for the visually impaired with features like e-book comprehension and image identification. Hearing impaired (HI) learners often suffer from inability to decipher speech intelligibility in noisy environments like classrooms, so deep neural network-based spectral mapping algorithms have been designed to disentangle reverberations for clean speech quality [77].

Various machine learning algorithms, explainable AI techniques (XAI), based on predictive analytics, as well as models employing artificial neural networks and Bayesian networks have been used for identifying learning difficulties in disabled students [87, 88]. Data for this purpose can be collected quantitatively and qualitatively through eye and lip tracking technology, electropherograms, and physical observation reports of learners and then used to create personalized learning strategies. Models like MALSEND [89] also correlate drop-out rates and success ratios in populations of disabled learners, allowing the correlation of underlying causes in a better way.

In a study of 163 participants which included both neuro-functional and neuro-impaired students, increased indicators (Torrance Figural Creativity Test) of creativity were found when collaborative AI was used for prompt generation and digital mind-mapping using the Mindomo tool [90]. A conversational agent, Woebot, has features like mood tracking, gamified delivery of cognitive behavior therapy based on decision tree with natural language inputs, which is inclusive of students facing mental health issues [91].

While positioning AI-based learning methods for long-term sustainability, one must consider cultural appropriation as an important factor in inclusivity. Learners belonging to all races, genders, ethnicities, colors, linguistic preferences should feel included in any technology deployed in the AI-assisted pedagogical processes [92].

Datasets used to train chatbots and facial recognition systems should be trained on diverse populations to make them "culturally competent" so that individual differences in beliefs, traditions, appearances are not discriminated against based on inherent bias in existing databases. If not taken care of, it could lead to a disconnect within minorities and marginalized communities in the AI learning environments.

4 Challenges to Sustaining the AI revolution

While the AI revolution in education has been progressing by leaps and bounds, it has been tough to identify and cope with challenges emerging from the shifting landscape of AIEd. While innovations and short-term perks are exciting in a cognitively rewarding educational setting, early detection of challenges is of paramount importance in sustaining the AI wave in education.

The collection and analysis of huge amounts of student data for use by AI-driven educational platforms [93–95] has been a concern for educators, parents, guardians, and administrators owing to the risks of sensitive data leaks rendering the impressionable learner community vulnerable to attacks by scammers, phishers, or even criminals. The confidentiality of data is constantly under threat owing to lack of resilience and quality of EdTech company frameworks in handling heavy traffic.

The ever-changing ecosystem of stakeholders and their dynamics with policymakers dictates the level of involvement in the decision-making processes pertaining to design and deployment of AI-driven educational technologies. Often EdTech companies employ coders and designers to create AI-powered educational products for commercialization that may not take account learner-specific sensibilities or adhere to ethical regulations [96].

Technical hallucinations are common in GenAI models which can lead to unreliability in generated information, simplistic content that is not up to standards, factual inaccuracies, and incomplete information rendering dependent educational stakeholders helpless in such situations. An interview-based case study by Nazaretsky et al. [97] with trained K12 science teachers demonstrates the tendency of teachers to feel uncertain or reject any recommendations that clash with their previous perceptions of pedagogical interventions, student performance, behavior, etc., in a classic diagnosis of Belief Perseverance and Confirmation Bias. What is more troubling is their expectation of AI to be an accurate know-it-all without keeping in mind the element of vagueness in assessing open-ended questions and the resulting confusion on confronting its limitations.

It has been seen that historical data used for training AI models can reinforce societal bias and skewed political beliefs. Imagine the disastrous consequences on the psyche of the educational stakeholders if the gender and racial discriminations trickle down to grading, admission procedures, recruitments as well as educational content recommendations in a supposedly fair and non-judgmental academic setup! AI in education is subject to algorithmic (learning, mapping, confirmation, etc.), user-related (interaction, stereotyping, exclusion, etc.), and data-related (historic, representation, and measurement) biases [98, 99].

These challenges can be overcome with careful planning and human interventions at all levels of educational planning for the future.

5 Sustainable Frameworks for the Future

Looking ahead, we need long-term solutions to tackle sustainability of the AI revolution in education and frameworks that are resilient against various challenges outlined in the previous section. Let us have a look at what aspects we need to examine while constructing such sustainable frameworks.

5.1 Avoiding Bias and Integrating Ethical Principles

While building models for long-term learning environments, systemic biases in existing data must be weeded out by diversification of data collection methods, inclusion of representative samples, and participatory contributions from different ethnicities, genders, and demographics [100]. Attention needs to be paid for creating culturally appropriate and politically neutral avatars in intelligent tutoring systems along with avoiding representational harm (stereotypical images of communities, racial slurs in text, etc.) [101]. An integral part of a sustainable framework would be understanding how collaborative dynamics in learning groups work across genders, spaces, and races. Learning analytics and patterns in discussions using approaches like Cross-MMLA (Multimodal Learning Analytics Across Spaces) [102] and Group Communication Analysis (GCA) [103, 104] could help evaluation of not only interdependent variables in communications but also help curate inclusive learning experiences across demographics.

AI-centered educational models should also be compatible with and designed in view of recommendations like the European Commission Ethics Guidelines for Trustworthy Artificial Intelligence [105] and latest metrics and standards on fair AI highlighted by Organisation for Economic Co-operation and Development (OECD) [106]. The pipelines for machine learning models in education could be periodically tested against fairness metrics and bias-mitigating algorithms within toolkits like IBM's AI Fairness 360 [107], Google Research developed What-If tool [108], and FairML [109]. Long-term diverse student datasets standardized from physical observations as well as ITS-based interactions like TIMSS, OULAD, EdNet, and CMU DataShop could be used in training sustainable models for keeping the bias to a bare minimum.

5.2 *Emphasis on AI-Integrated Teacher Education*

Frameworks including AIEd in teacher education curricula will help to unleash the full potential to improve educational outcomes and enrich the teaching–learning experience for educators as well as students. Research studies have pointed at improvement of teacher–student interactions, enhancement of learning outcomes, tailored classroom experiences, and better performances of students in settings with high teacher competencies in using AI, thereby highlighting the importance of teacher training [110–116].

However, the training of existing teachers is still a valid question. Mass and proper training programs are required for in-service teachers to update their knowledge from time to time, Frameworks like Technological Pedagogical and Content Knowledge (TPACK) can be useful to reinforce their AI-integrated learning curves. A study conducted on 216 teachers undergoing AI- blended professional development program [117] reported that teachers felt that the AI-based training introduced different methodologies than their traditional teaching practices and that they experienced challenges to their Teacher Professional Identity (TPI). Treating the combination of Human Intelligence and Artificial Intelligence as navigators, collaborators and innovators, researchers found that these joint roles could not only drive creativity in the classroom but also promote collaborative practices through engagement and discussion through teacher working groups.

With the launch of initiatives like AI4All and introduction of AI subjects in school and college curriculum, capacity building programs at the pre-service and different in-service stages are a must for teachers. These can provide continuous and real-time feedback loops that are essential to identify bottlenecks in ongoing AI literacy programs [118].

5.3 Building Robust Policy Frameworks

The Beijing consensus is a major landmark in AIEd-based policymaking as it marks the creation of the first document designed by UNESCO to guide frameworks based on benefit-risk analysis by suggesting best practices for integrating AI skills in ICT-competency programs for achieving SDG4 by 2030 [119]. The document proposes several recommendations for policymaking such as interdisciplinary, inter-sectoral cooperation of neuroscience experts, psychologists, government representatives, educationists, students, developers, corporate hierarchies, for policy planning in AIEd, as well as facilitating pilot evaluations of AI-based educational tools at multiple levels, evidence-based approaches for human–machine collaborations, and realigning curriculum periodically to reflect learner needs coupled with employability demands.

These guidelines have been examined in a research by Chan et al. [120] that proposes an AI Ecological Education Policy Framework based on joint perspectives of 457 participants from Hong Kong universities that focuses on three pillars of interest—Operation, Governance, and Pedagogy. A significant discussion relevant for AIEd policymakers is centered around the robust guidelines required to define transparency, accountability, fallacies, plagiarism, ethical dilemmas, security, and data usage.

Based on a study conducted on building teacher trust around AI-driven educational platforms [115], a major recommendation for policymakers comes through that focuses on how AI-EdTech should be flexible for customizing to classroom scenarios and that teachers should have the authority to modify/override AI-suggested learning paths, assessments, and reports.

5.4 Resilience Against Attacks

Security is of paramount importance to avoid misuse of AI technologies and resilient frameworks must be established across educational institutions through strengthening of ICT infrastructure, encryption, awareness session with all stakeholders as well as recruitment of ICT security officers to protect sensitive information. A multifold security model has been proposed [121] based on AI techniques to prevent data theft, financial fraud, access control in educational governance where security specialists, technology managers, institutional heads, teachers, students, and staff play key roles in upholding the highest standards of privacy protection.

Detection of unauthorized access by facial recognition, analysis of threatening patterns in emails and chats, and user behavior through machine learning pathways and intrusion-detection system based on deep-learning analysis of suspicious network traffic are some of the techniques that can be deployed for protective measures for building a sustainable framework for security that is resilient to cyberattacks.

6 Conclusion

In this chapter, we have explored how artificial intelligence techniques have been applied to personalize instructional strategies, plan and create lessons, use learner analytics for improved performances, and improvise assessments while aiding educational governance and building engaging learning environments. As sustainable solutions aim at equitable access for all, the use of artificial intelligence in education demonstrates strong capabilities for building frameworks and applications aimed at integrating those excluded by socio-cultural, linguistic, geographical as well as physical/mental disabilities within the mainstream learning environments. Systemic removal as well as prevention of any bias in the classroom and beyond forms the core of human-centered AIEd. Heterogeneous bias free datasets could be shared across borders for training AI algorithms for cultural inclusivity. The current status of AIEd advocates for more transparent and flexible policies instead of the previous black-box approaches.

Sustaining AIEd would be difficult without human expertise and vision of teachers, and a framework that supports relevant AI training in pre-service as well as in-service stages with spaces for constant dialog and collaborative efforts would be highly desirable. Sustainable educational frameworks and policies should be designed in collaboration with academicians, researchers, and developers.

To conclude, as AIEd grows in dimensions with each passing day, we must realize that at the end of the day, it is not a panacea and that there can be no substitute for the creative and original ideas for learning strategies devised by human teachers, and the personal empathetic viewpoints of educators that shape the destiny of students. Thereby, our focus needs to stay on utilizing the exciting features of AI and mastering it but not depending on it wholly, to sustain these technologies for the future generation.

Acknowledgments The authors acknowledge the infrastructural facilities at Department of Education, IIT Kharagpur

Competing Interests The authors hereby declare that there are no competing interests regarding the publication of this chapter.

References

1. Leal Filho W, Ribeiro PCC, Mazutti J, et al (2024) Using artificial intelligence to implement the UN sustainable development goals at higher education institutions. Int J Sustain Dev World Ecol 31:726–745. doi:https://doi.org/10.1080/13504509.2024.2327584
2. Kamruzzaman MM, Alanazi S, Alruwaili M, et al (2023) AI-and IoT-assisted sustainable education systems during pandemics, such as COVID-19, for smart cities. Sustainability 15:8354
3. Yuskovych-Zhukovska V, Poplavska T, Diachenko O, et al (2022) Application of artificial intelligence in education. Problems and opportunities for sustainable development. BRAIN Broad Res Artif Intell Neurosci 13:339–356
4. Goralski MA, Tan TK (2020) Artificial intelligence and sustainable development. Int J Manag Educ 18:100330
5. Klašnja-Milićević A, Ivanović M (2021) E-learning personalization systems and sustainable education
6. Khan S, Mazhar T, Shahzad T, et al (2025) Harnessing AI for sustainable higher education: ethical considerations, operational efficiency, and future directions. Discov Sustain 6:23. doi:https://doi.org/10.1007/s43621-025-00809-6
7. Jagannathan S (2021) Reimagining digital learning for sustainable development: how upskilling, data analytics, and educational technologies close the skills gap. Routledge
8. Chiu TK, Chai C (2020) Sustainable curriculum planning for artificial intelligence education: A self-determination theory perspective. Sustainability 12:5568
9. Hui ESYE (2024) Incorporating Bloom's taxonomy into promoting cognitive thinking mechanism in artificial intelligence-supported learning environments. Interact Learn Environ 1–14. doi:https://doi.org/10.1080/10494820.2024.2364237
10. Gonsalves C (2024) Generative AI's Impact on Critical Thinking: Revisiting Bloom's Taxonomy. J Mark Educ 02734753241305980. doi:https://doi.org/10.1177/02734753241305980
11. Bai S, Lo CK, Yang C (2024) Enhancing instructional design learning: a comparative study of scaffolding by a 5E instructional model-informed artificial intelligence chatbot and a human teacher. Interact Learn Environ 1–20. doi:https://doi.org/10.1080/10494820.2024.2420184
12. Goodman J, Handa V, Wilson RE, Bradbury LU (2024) Promises and pitfalls: Using an AI chatbot as a tool in 5E lesson planning. Innov Sci Teach Educ 9:1–13
13. Mohammadi L, Asadi M, Taheri R (2024) Transforming EFL Lesson Planning with To Teach AI: Insights from Teachers Perspectives. Technol Assist Lang Educ 2. https://doi.org/10.22126/tale.2025.11490.1080
14. Pender H-L, Bohl L, Schönberger M, Knopf J (2022) An AI-based lesson planning software to support competence-based learning. In: 8th International Conference on Higher Education Advances (HEAd'22). Editorial Universitat Politècnica de València, pp 1033–1041
15. Ranawat R, Venkataraman A, Subramanian L (2021) CollectiveTeach: A System To Generate And Sequence Web-Annotated Lesson Plans. In: ACM SIGCAS Conference on Computing and Sustainable Societies (COMPASS). ACM, Virtual Event Australia, pp. 1–13
16. Karaman MR (2024) Are Lesson Plans Created by ChatGPT More Effective? An Experimental Study. Int J Technol Educ 7:107–127
17. van den Berg G, du Plessis E (2023) ChatGPT and generative AI: Possibilities for its contribution to lesson planning, critical thinking and openness in teacher education. Educ Sci 13:998
18. Park HK, Park J (2024) Lesson Plan Development for Pre-Service Elementary School Teachers through Collaboration with AI: Focused on ChatGPT
19. Demir G, Cimen E (2024) Lesson Plan Preparation Process with ChatGPT in Mathematics Teaching: An Example of Research and Practice. Technol Innov Spec Educ Res 4:83–98
20. Kalenda PJ, Rath L, Abugasea Heidt M, Wright A (2025) Pre-service Teacher Perceptions of ChatGPT for Lesson Plan Generation. J Educ Technol Syst 53:219–241. doi:https://doi.org/10.1177/00472395241301388

21. Cameron S, Mesiti C (2024) What Kind of Mathematics Teacher Is ChatGPT? Identifying the Pedagogical Practices Preferenced by Generative AI Tools When Preparing Lesson Plans. Math Educ Res Group Australas
22. Lee G-G, Zhai X (2024) Using ChatGPT for Science Learning: A Study on Pre-service Teachers' Lesson Planning. IEEE Trans Learn Technol
23. S DDAO Manochandar, Kuchipudi DKR Jyothsna Devi, Chotemiya DBG Mr Mohmmad, Sv DABA Ramanan (2024) AI & CHATGPT TOOLS FOR TEACHING LEARNING PROCESS. REDSHINE Publication
24. Sundari MS, Penthala HR, Nayyar A (2024) Transforming Education through AI-Enhanced Content Creation and Personalized Learning Experiences. In: Impact of Artificial Intelligence on Society. Chapman and Hall/CRC
25. Gligorea I, Cioca M, Oancea R, et al (2023) Adaptive Learning Using Artificial Intelligence in e-Learning: A Literature Review. Educ Sci 13:1216. doi:https://doi.org/10.3390/educsci13 121216
26. Kim Y, Soyata T, Behnagh RF (2018) Towards Emotionally Aware AI Smart Classroom: Current Issues and Directions for Engineering and Education. IEEE Access 6:5308–5331. doi:https://doi.org/10.1109/ACCESS.2018.2791861
27. Wei X, Wang L, Lee L-K, Liu R (2024) Multiple Generative AI Pedagogical Agents in Augmented Reality Environments: A Study on Implementing the 5E Model in Science Education. J Educ Comput Res 07356331241305519. doi:https://doi.org/10.1177/073563312413 05519
28. Tao R, Tian J, Liang Y (2024) Virtual Simulation Aids for AI Curriculum Learning: A Quantitative Evaluation with the 5E Instructional Model. In: 2024 International Symposium on Educational Technology (ISET). pp 119–123
29. VR AL You Calligrapher|Meta Quest Game. https://www.altlabvr.com/you-calligrapher. Accessed 12 Mar 2025
30. Łysakowski M, Gapsa J, Lyu C, et al (2025) Enhancing augmented reality with machine learning for hands-on origami training. Front Virtual Real 6:1499830
31. Torshizi PG, Tarashiyoun H, Moghaddam M, Marsella S (2024) Towards an Intelligent Augmented Reality Tutoring System for Precision Manufacturing. In: Proceedings of the 18th International Conference of the Learning Sciences-ICLS 2024, pp. 1578-1581. International Society of the Learning Sciences
32. Joshi S, Agbo FJ, Jormanainen I (2023) Towards Enhancing Children's Science Education using Augmented Reality and Computer Vision. In: 2023 IEEE Global Engineering Education Conference (EDUCON). IEEE, pp 1–3
33. Kim J, Shim J (2022) Development of an AR-Based AI Education App for Non-Majors. IEEE Access 10:14149–14156. doi:https://doi.org/10.1109/ACCESS.2022.3145355
34. Song Y, Wu K, Ding J (2024) Developing an immersive game-based learning platform with generative artificial intelligence and virtual reality technologies–"LearningverseVR." Comput Educ X Real 4:100069
35. Tseng S-S, Chen S-N, Yang T-Y (2022) Building an AR-based smart campus platform. Multimed Tools Appl 81:5695–5716. doi:https://doi.org/10.1007/s11042-021-11702-5
36. El Gourari A, Ezzini M, Raoufi M, et al (2024) Development of an Intelligent Communication Platform for Remote Practical Work in Education: Integrating AR, VR, and Cloud Technologies. In: ITM Web of Conferences. EDP Sciences, p 04013
37. Owan VJ, Abang KB, Idika DO, et al (2023) Exploring the potential of artificial intelligence tools in educational measurement and assessment. Eurasia J Math Sci Technol Educ 19:em2307. https://doi.org/10.29333/ejmste/13428
38. Fitria TN (2021) Grammarly as AI-powered English writing assistant: Students' alternative for writing English. Metathesis J Engl Lang Lit Teach 5:65–78
39. Fonseca I, Martins NC, Lopes F (2024) Exploring interactive tools for education and online assessment: a comparative analysis. In: 2024 21st International Conference on Information Technology Based Higher Education and Training (ITHET). IEEE, pp 1–6

40. Gan W, Qi Z, Wu J, Lin JC-W (2023) Large language models in education: Vision and opportunities. In: 2023 IEEE international conference on big data (BigData). IEEE, pp 4776–4785
41. Xu H, Gan W, Qi Z, et al (2024) Large Language Models for Education: A Survey
42. Darvishi A, Khosravi H, Sadiq S, Gašević D (2022) Incorporating AI and learning analytics to build trustworthy peer assessment systems. Br J Educ Technol 53:844–875. doi:https://doi.org/10.1111/bjet.13233
43. Abdallah AK, Abdallah RK (2025) Smart Solutions for Smarter Schools: Leveraging Artificial Intelligence to Revolutionize Educational Administration and Leadership. In: Encyclopedia of Information Science and Technology, Sixth Edition. IGI Global Scientific Publishing, pp. 1–14
44. Wattanapanit N, Tanchaisak K, Niyomves B, Kenaphoom S (2024) An Educational Administration Innovation in the Digital Age and Sustainable Development. In: Modern Management Science Practices in the Age of AI. IGI Global Scientific Publishing, pp. 211–236
45. Mupaikwa E (2025) The Application of Artificial Intelligence in Educational Administration. In: AI Adoption and Diffusion in Education. IGI Global Scientific Publishing, pp. 209–230
46. Sorensen LC (2019) "Big Data" in Educational Administration: An Application for Predicting School Dropout Risk. Educ Adm Q 55:404–446. doi:https://doi.org/10.1177/0013161X18799439
47. Kawamura R, Shirai S, Takemura N, et al (2021) Detecting Drowsy Learners at the Wheel of e-Learning Platforms With Multimodal Learning Analytics. IEEE Access 9:115165–115174. doi:https://doi.org/10.1109/ACCESS.2021.3104805
48. Wang Y (2021) Artificial intelligence in educational leadership: a symbiotic role of human-artificial intelligence decision-making. J Educ Adm 59:256–270
49. Tyson MM, Sauers NJ (2021) School leaders' adoption and implementation of artificial intelligence. J Educ Adm 59:271–285
50. Mutambik I (2024) The Use of AI-Driven Automation to Enhance Student Learning Experiences in the KSA: An Alternative Pathway to Sustainable Education. Sustainability 16:5970. doi:https://doi.org/10.3390/su16145970
51. Dwivedi A, Dsouza M, Aggarwal A (2024) AI-Driven School Management System: A React-Based Web Application Enhancing Educational Administration and Student Performance Analytics. Int J Innov Sci Res Technol IJISRT 3042–3054. https://doi.org/10.38124/ijisrt/IJISRT24APR2465
52. Afrin S, Roksana S, Akram R (2025) AI-Enhanced Robotic Process Automation: A Review of Intelligent Automation Innovations. IEEE Access 13:173–197. doi:https://doi.org/10.1109/ACCESS.2024.3513279
53. Chang DH, Lin MP-C, Hajian S, Wang QQ (2023) Educational Design Principles of Using AI Chatbot That Supports Self-Regulated Learning in Education: Goal Setting, Feedback, and Personalization. Sustainability 15:12921. doi:https://doi.org/10.3390/su151712921
54. Liu L, Subbareddy R, Raghavendra CG (2022) AI Intelligence Chatbot to Improve Students Learning in the Higher Education Platform. J Interconnect Netw 22:2143032. doi:https://doi.org/10.1142/S0219265921430325
55. Menkhoff T, Gan B (2023) Engaging students through conversational chatbots and digital content: A climate action perspective.
56. Park A, Lee SB (2024) Examining AI and Systemic Factors for Improved Chatbot Sustainability. J Comput Inf Syst 64:728–742. doi:https://doi.org/10.1080/08874417.2023.2251416
57. Endurance A, Onah Eunice N, Uzoma Anthonia C, Andor Sebastine E (2021) Integration of Artificial Intelligence Tool (Ai-Chatbot) into Teaching and Learning: A Panacea for Improving UniversitiesEducational and Administrative Duties in South-South, Nigeria, J Comput Sci Syst Biol 14:357
58. Ait Baha T, El Hajji M, Es-Saady Y, Fadili H (2024) The impact of educational chatbot on student learning experience. Educ Inf Technol 29:10153–10176. doi:https://doi.org/10.1007/s10639-023-12166-w

59. Deng X, Yu Z (2023) A meta-analysis and systematic review of the effect of chatbot technology use in sustainable education. Sustainability 15:2940
60. Dai W, Lin J, Jin H, et al (2023) Can large language models provide feedback to students? A case study on ChatGPT. In: 2023 IEEE International Conference on Advanced Learning Technologies (ICALT). IEEE, pp 323–325
61. Al-Sharafi MA, Al-Emran M, Iranmanesh M, et al (2023) Understanding the impact of knowledge management factors on the sustainable use of AI-based chatbots for educational purposes using a hybrid SEM-ANN approach. Interact Learn Environ 31:7491–7510. doi:https://doi.org/10.1080/10494820.2022.2075014
62. Popescu RI, Sabie OM, Trușcă MI (2023) The Contribution of Artificial Intelligence to Stimulating the Innovation of Educational Services and University Programs in Public Administration. Transylv Rev Adm Sci 19:85–108
63. Mousavinasab E, Zarifsanaiey N, R. Niakan Kalhori, Rakhshan M, Keikha L, Ghazi Saeedi M et al (2021) Intelligent tutoring systems: a systematic review of characteristics, applications, and evaluation methods. Interact Learn Environ 29:142–163. doi:https://doi.org/10.1080/10494820.2018.1558257
64. Keleş A, Ocak R, Keleş A, Gülcü A (2009) ZOSMAT: Web-based intelligent tutoring system for teaching–learning process. Expert Syst Appl 36:1229–1239
65. Kawamura R, Shirai S, Takemura N, et al (2021) Detecting Drowsy Learners at the Wheel of e-Learning Platforms With Multimodal Learning Analytics. IEEE Access 9:115165–115174. doi:https://doi.org/10.1109/ACCESS.2021.3104805
66. Keleş A, Ocak R, Keleş A, Gülcü A (2009) ZOSMAT: Web-based intelligent tutoring system for teaching–learning process. Expert Syst Appl 36:1229–1239
67. Myneni LS, Narayanan NH, Rebello S, et al (2013) An Interactive and Intelligent Learning System for Physics Education. IEEE Trans Learn Technol 6:228–239. doi:https://doi.org/10.1109/TLT.2013.26
68. Mostafavi B, Barnes T (2017) Evolution of an Intelligent Deductive Logic Tutor Using Data-Driven Elements. Int J Artif Intell Educ 27:5–36. doi:https://doi.org/10.1007/s40593-016-0112-1
69. Soh L-K, Khandaker N, Jiang H I-MINDS: A Multiagent System for Intelligent Computer-Supported Collaborative Learning and Classroom Management
70. Craig SD, Hu X, Graesser AC, et al (2013) The impact of a technology-based mathematics after-school program using ALEKS on student's knowledge and behaviors. Comput Educ 68:495–504. doi:https://doi.org/10.1016/j.compedu.2013.06.010
71. Isaza Domínguez LG, Velasquez Clavijo F, Robles-Gómez A, Pastor-Vargas R (2024) A Sustainable Educational Tool for Engineering Education Based on Learning Styles, AI, and Neural Networks Aligning with the UN 2030 Agenda for Sustainable Development. Sustainability 16:8923. doi:https://doi.org/10.3390/su16208923
72. Ahmad FK (2015) Use of Assistive Technology in Inclusive Education: Making Room for Diverse Learning Needs. 6:
73. Mitre X, Zeneli M (2024) Using AI to Improve Accessibility and Inclusivity in Higher Education for Students with Disabilities. In: 2024 21st International Conference on Information Technology Based Higher Education and Training (ITHET). pp 1–8
74. Inc C The UDL Guidelines. https://udlguidelines.cast.org. Accessed 19 Feb 2025
75. Web Content Accessibility Guidelines (WCAG) 2.0. https://www.w3.org/TR/WCAG20/. Accessed 19 Feb 2025
76. Felix SM, Kumar S, Veeramuthu A (2018) A smart personal AI assistant for visually impaired people. In: 2018 2nd international conference on trends in electronics and informatics (ICOEI). IEEE, pp 1245–1250
77. Zhao Y, Wang D, Merks I, Zhang T (2016) DNN-based enhancement of noisy and reverberant speech. In: 2016 IEEE International Conference on Acoustics, Speech and Signal Processing (ICASSP). pp 6525–6529
78. Granquist C, Sun SY, Montezuma SR, et al (2021) Evaluation and Comparison of Artificial Intelligence Vision Aids: Orcam MyEye 1 and Seeing AI. J Vis Impair Blind 115:277–285. doi:https://doi.org/10.1177/0145482X211027492

79. Aicha Z, Asmae Z, Hamdaoui R, et al (2024) Real-Time Implementation of an AI-Based Virtual Sign Language Recognition and Interpretation System
80. Professional & AI-Based Captions for Deaf & HoH|Ava. https://www.ava.me/. Accessed 19 Feb 2025
81. Seebun GR, Nagowah L (2020) Let's Talk: An Assistive Mobile Technology for Hearing and Speech Impaired Persons. In: 2020 3rd International Conference on Emerging Trends in Electrical, Electronic and Communications Engineering (ELECOM). pp 210–215
82. Logan® ProxTalker® Mid-Tech AAC Device–LoganTech. https://logantech.com/products/proxtalker. Accessed 19 Feb 2025
83. Pak NS, Bailey KM, Ledford JR, Kaiser AP (2023) Comparing Interventions With Speech-Generating Devices and Other Augmentative and Alternative Communication Modes: A Meta-Analysis. Am J Speech Lang Pathol 32:786–802. doi:https://doi.org/10.1044/2022_AJSLP-22-00220
84. Asvestopoulou T, Manousaki V, Psistakis A, et al (2019) DysLexML: Screening Tool for Dyslexia Using Machine Learning
85. Rajapakse S, Polwattage D, Guruge U, et al (2018) ALEXZA: A mobile application for dyslexics utilizing artificial intelligence and machine learning concepts. In: 2018 3rd International Conference on Information Technology Research (ICITR). IEEE, pp 1–6
86. Lister K, Coughlan T, Kenny I, et al (2021) Taylor, the disability disclosure virtual assistant: A case study of participatory research with disabled students. Educ Sci 11:587
87. Sharif MS, Elmedany W (2022) A proposed machine learning based approach to support students with learning difficulties in the post-pandemic norm. In: 2022 IEEE Global Engineering Education Conference (EDUCON). IEEE, pp 1988–1993
88. Raji NR, Kumar RMS, Biji CL (2024) Explainable Machine Learning Prediction for the Academic Performance of Deaf Scholars. IEEE Access
89. Sobnath D, Isiaq SO, Rehman IU, Nasralla MM (2020) Using Machine Learning Advances to Unravel Patterns in Subject Areas and Performances of University Students with Special Educational Needs and Disabilities (MALSEND): A Conceptual Approach. In: Yang X-S, Sherratt S, Dey N, Joshi A (eds) Fourth International Congress on Information and Communication Technology. Springer Singapore, Singapore, pp. 509–517
90. Fang M, Abdallah AK, Vorfolomeyeva O (2024) Collaborative AI-enhanced digital mind-mapping as a tool for stimulating creative thinking in inclusive education for students with neurodevelopmental disorders. BMC Psychol 12:488. doi:https://doi.org/10.1186/s40359-024-01975-4
91. Fitzpatrick KK, Darcy A, Vierhile M (2017) Delivering Cognitive Behavior Therapy to Young Adults With Symptoms of Depression and Anxiety Using a Fully Automated Conversational Agent (Woebot): A Randomized Controlled Trial. JMIR Ment Health 201742e19 https://mental.jmir.org/2017/2/e19/. https://doi.org/10.2196/mental.7785
92. Kazimzade G, Patzer Y, Pinkwart N (2019) Artificial Intelligence in Education Meets Inclusive Educational Technology—The Technical State-of-the-Art and Possible Directions
93. Huang L (2023) Ethics of artificial intelligence in education: Student privacy and data protection. Sci Insights Educ Front 16:2577–2587
94. Vaza RN, Parmar AB, Mishra PS, et al (2024) Security And Privacy Concerns In AI-Enabled Iot Educational Frameworks: An In-Depth Analysis. Educ Adm Theory Pract 30:8436–8445
95. Kamenskih A (2022) The analysis of security and privacy risks in smart education environments. J Smart Cities Soc 1:17–29
96. Buckingham Shum SJ, Luckin R (2019) Learning analytics and AI: Politics, pedagogy and practices. Br J Educ Technol 50:2785–2793. doi:https://doi.org/10.1111/bjet.12880
97. Nazaretsky T, Cukurova M, Ariely M, Alexandron G (2021) Confirmation bias and trust: Human factors that influence teachers' attitudes towards AI-based educational technology
98. Chinta SV, Wang Z, Yin Z, et al (2024) FairAIED: Navigating Fairness, Bias, and Ethics in Educational AI Applications
99. Baker RS, Hawn A (2022) Algorithmic Bias in Education. Int J Artif Intell Educ 32:1052–1092. doi:https://doi.org/10.1007/s40593-021-00285-9

100. Roscoe RD (2021) Designing for diversity: Inclusive sampling. Ergodesign HCI 9:
101. Mayfield E, Madaio M, Prabhumoye S, et al (2019) Equity beyond bias in language technologies for education. In: Proceedings of the fourteenth workshop on innovative use of NLP for building educational applications. pp. 444–460
102. Spikol D, Lorenzo MLB, Cukurova M, et al (2022) CROSSMMLA & SLE Workshop: Learning Analytics for Smart Learning Environments Crossing Physical and Virtual Learning Spaces. Companion Proc 12th 173
103. Dowell NM, Lin Y, Godfrey A, Brooks C (2020) Exploring the Relationship between Emergent Sociocognitive Roles, Collaborative Problem-Solving Skills, and Outcomes: A Group Communication Analysis. J Learn Anal 7:38–57
104. Dowell NMM, Nixon TM, Graesser AC (2019) Group communication analysis: A computational linguistics approach for detecting sociocognitive roles in multiparty interactions. Behav Res Methods 51:1007–1041. doi:https://doi.org/10.3758/s13428-018-1102-z
105. Ethics guidelines for trustworthy AI|Shaping Europe's digital future. https://digital-strategy.ec.europa.eu/en/library/ethics-guidelines-trustworthy-ai. Accessed 20 Feb 2025
106. The OECD Artificial Intelligence Policy Observatory-OECD.AI. https://oecd.ai/en/. Accessed 20 Feb 2025
107. Home. In: AI Fairness 360. https://ai-fairness-360.org/. Accessed 20 Feb 2025
108. The What-If Tool: Interactive Probing of Machine Learning Models. https://research.google/pubs/the-what-if-tool-interactive-probing-of-machine-learning-models/. Accessed 20 Feb 2025
109. Adebayo JA (2016) FairML: ToolBox for diagnosing bias in predictive modeling. Thesis, Massachusetts Institute of Technology
110. AI in teacher education: Unlocking new dimensions in teaching support, inclusive learning, and digital literacy—Zhang—2024—Journal of Computer Assisted Learning-Wiley Online Library. https://onlinelibrary.wiley.com/doi/10.1111/jcal.12988. Accessed 20 Feb 2025
111. Henze J, Schatz C, Malik S, Bresges A (2022) How might we raise interest in robotics, coding, artificial intelligence, STEAM and sustainable development in university and on-the-job teacher training? In: Frontiers in education. Frontiers Media SA, p 872637
112. Sperling K, Stenberg C-J, McGrath C, et al (2024) In search of artificial intelligence (AI) literacy in Teacher Education: A scoping review. Comput Educ Open 100169
113. Kim J (2024) Leading teachers' perspective on teacher-AI collaboration in education. Educ Inf Technol 29:8693–8724. doi:https://doi.org/10.1007/s10639-023-12109-5
114. Chen S, Sun Y, Zhang H, Liu Q (2021) Role of Teacher Training Based on School-enterprise Cooperation in Artificial Intelligence Specialty. In: 2021 International Symposium on Advances in Informatics, Electronics and Education (ISAIEE). IEEE, pp. 256–259
115. Nazaretsky T, Ariely M, Cukurova M, Alexandron G (2022) Teachers' trust in AI—powered educational technology and a professional development program to improve it. Br J Educ Technol 53:914–931. doi:https://doi.org/10.1111/bjet.13232
116. Seo K, Tang J, Roll I, et al (2021) The impact of artificial intelligence on learner–instructor interaction in online learning. Int J Educ Technol High Educ 18:54. doi:https://doi.org/10.1186/s41239-021-00292-9
117. Through tensions to identity-based motivations: Exploring teacher professional identity in Artificial Intelligence-enhanced teacher training—ScienceDirect. https://www.sciencedirect.com/science/article/pii/S0742051X24002683#sec2. Accessed 20 Feb 2025
118. Vazhayil A, Shetty R, Bhavani RR, Akshay N (2019) Focusing on Teacher Education to Introduce AI in Schools: Perspectives and Illustrative Findings. In: 2019 IEEE Tenth International Conference on Technology for Education (T4E). pp 71–77
119. AI and education: guidance for policy-makers - UNESCO Digital Library. https://unesdoc.unesco.org/ark:/48223/pf0000376709?locale=en. Accessed 20 Feb 2025
120. Chan CKY (2023) A comprehensive AI policy education framework for university teaching and learning. Int J Educ Technol High Educ 20:38. doi:https://doi.org/10.1186/s41239-023-00408-3

121. Toapanta SMT, Del Pozo Durango RH, Díaz EZG, et al (2023) Proposal for a security model applying artificial intelligence for administrative management in a higher education institution. In: 2023 International Conference on Computer, Information and Telecommunication Systems (CITS). pp. 1–5

Correction to: AI-Driven Quality Inspection for Sustainable Manufacturing in Industry 4.0

K. Indra Gandhi, P. K. Jawahar, G. Kannan, Abdul Rajak A. R., and Ashutosh Mishra

Correction to: Chapter 5 in: A. Khanna et al. (eds.), *Artificial Intelligence for Sustainable Development Goals*, https://doi.org/10.1007/978-3-032-28217-0_5

The book was inadvertently published with an incorrect affiliation for two of the Authors of Chapter 5

For this book, affiliation for Author in Chapter 5 needs to be updated:

In reference to Chapter 5:

These two affiliations need to be correct:

A. R. Abdul Rajak (abdulrazak@dubai.bits-pilani.ac.in)

Ashutosh Mishra (ashutosh@dubai.bits-pilani.ac.in)

Birla Institute of Technology and Science, Pilani, Dubai, United Arab Emirates.

The updated version of this chapter can be found at https://doi.org/10.1007/978-3-032-28217-0_5

A. Khanna et al. (eds.), *Artificial Intelligence for Sustainable Development Goals*, https://doi.org/10.1007/978-3-032-28217-0_12

MIX
Papier aus verantwortungsvollen Quellen
Paper from responsible sources
FSC® C105338

If you have any concerns about our products, you can contact us on
ProductSafety@springernature.com

In case Publisher is established outside the EU, the EU authorized representative is:
Springer Nature Customer Service Center GmbH
Europaplatz 3, 69115 Heidelberg, Germany

Printed by Libri Plureos GmbH
in Hamburg, Germany